Quantitative Analy

Ron Jones

Pitman Publishing
128 Long Acre, London WC2E 9AN

A Division of Longman Group UK Ltd.

© Longman Group UK Ltd 1990

British Library Cataloguing in Publication Data

Jones, Ron, *1939–*
   Quantitative analysis.
    1. Accounting. Quantitative methods. I. Title
    I. Title
657

ISBN 0 273 02998 3

Typeset, printed and bound in Great Britain

Contents

Preface

This book has been written primarily to cover the ACCA Paper 2.6, Quantitative Analysis, though it should also be of use to other post-Foundation Level students. A certain level of knowledge is assumed, and in particular you will need to be familiar with statistical measures of central tendency and dispersion, and with compounding and discounting techniques. You will also find the section on allocation easier to follow if you are familiar with matrix arithmetic, though a knowledge of matrices is not essential.

At the end of each chapter there is a set of review questions for you to attempt. The purpose of these questions is to review your understanding of the chapter and should not be seen as typical of the level of difficulty you would expect to meet in a Second Level examination. At the end of the book you will find a selection of examination questions. A full answer is given to each question.

Ron Jones

Part One Probability and expectation

1 Probability

1.1 Introduction

You will probably be aware that much of a statistician's time is spent measuring data and drawing conclusions based on the measurements. Sometimes, all of the data is available to the statistician and providing care is taken, there is no reason why any conclusions drawn on the measurements taken should not be very accurate. In such cases, the statistician is said to have perfect knowledge of the population being investigated. Unfortunately, this will not be the usual situation. In most cases, the statistician will not have all the details about the entire population, and will be unable to collect this information because of the cost involved. To take an example, suppose the Department of Health in a certain country wishes to know the average weight of adult males in order to monitor levels of obesity. It would not be possible to measure the weight of every male, so instead the statistician would have to make do with a sample. Now if the sample is carefully drawn, and if the sample is not too small, then the sample average will give a good approximation to the population average. However, because the entire population has not been examined, the statistician can never be completely sure of any conclusions made and will quote a degree of confidence or level of probability for his results. So, in order to understand the principles of sampling, we must first understand the meaning and theory of probability.

1.2 Some definitions

Let us suppose that we toss a coin. This experiment can have two outcomes: the coin can land with either heads showing or tails showing. The result of such an experiment is called an *event*. Notice that in this case the events are *mutually exclusive*, that is, if a head occurs, then a tail cannot occur at the same time. Now not all events are mutually exclusive, and we must be careful to recognise when they are mutually exclusive and when they are not. Suppose, for example, we are selecting a card from a pack; and the first event is that the card is red and the second event is that the card is an ace. If we draw the ace of hearts or the ace of diamonds, then both events have occurred simultaneously. Clearly the events are *not* mutually exclusive. Usually, we use a capital E with a suitable subscript to identify the events in an experiment. We could write the events in the coin-spinning experiment like this:

E_1 = the coin shows a head
E_2 = the coin shows a tail

Notice too that E_1 and E_2 are *collectively exhaustive*: that is, they account for all the logical possibilities (we treat with contempt the suggestion that the coin lands on its edge!).

Sometimes we will find it convenient to consider the situations when the event would not occur, and we can do this by placing a dash after the symbol for the event, so we could write:

E_1' = the coin does not show a head

Can you see that in this case E_1' and E_2 are equivalent events? If the coin does not show a head, then it must show a tail. We conclude that if $E_1' = E_2$, then E_1 and E_2 must both be mutually exclusive and collectively exhaustive.

1.3 How we measure probability

Let us consider a certain experiment, and list all the possible events; that is, we will ensure that the events are collectively exhaustive. Events that are equally likely will be assigned the same 'weighting'. Suppose, for example, we again consider tossing a coin. We have:

E_1 = the coin shows a head; E_2 = the coin shows a tail

If we make the assumption that the coin is unbiased, then E_1 and E_2 must be equally likely. What we will now do is to assign 'weights' to the events in proportion to the likelihood that they will occur. So we have:

	Weight
E_1	1
E_2	1

Now the probability that the event E occurs is

$$P(E) = \frac{\text{Weighting for event } E}{\text{Sum of the weights}}$$

So $P(E_1) = \frac{1}{2}$ (or 0.5), and $P(E_2) = \frac{1}{2}$.

So we can see that the probability of obtaining a head with a single throw of a coin is one half. Just what do we mean by this? Well, it is obvious that we cannot demonstrate probability with a single toss of a coin. In fact, the outcome of a single toss depends on the force we exert together with the way the coin was originally facing. It has nothing to do with probability! When we state that the probability of a head is a half, we are surely making some prediction as to the proportion of heads occurring if we repeat this experiment many times. The outcome of each *individual* toss is determined by the forces mentioned earlier, but the outcome of many tosses obeys a law of 'mass behaviour', and it is this 'mass behaviour' that our probability measure is trying to predict. So when we state the probability of a head is a half, we mean that if the coin is tossed a large number of times, then we would expect heads to occur on 50 per cent of occasions.

Let us now see if we can restate our measure of probability. If an experiment has N *equally likely* outcomes, n of which constitute event E, then we can state that

$$P(E) = \frac{n}{N}$$

Suppose, then, we wished to find the probability of drawing an ace from a well-shuffled pack of cards. Here we have $N = 52$ (there are 52 cards in a pack, all of them have an equal chance of being drawn) and $n = 4$, so $P(\text{ace}) = \frac{4}{52} = \frac{1}{13}$. Easy, isn't it?

Now suppose that $n = N$. This implies that each and every outcome must constitute event E. In other words, if we perform the experiment we are absolutely certain that event E will occur. Moreover, if $n = N$, then $P(E) = 1$, so we assign a probability measure of 1 to events that are absolutely certain to occur. If the event E cannot possibly occur, then n = 0 and $P(E) = 0$, so we assign a zero probability measure to events that are absolutely impossible. As absolute certainty and absolute impossibility are at opposite ends of the spectrum, we have now obtained limits to our measure of probability – $P(E)$ must lie between zero and one. We can write this statement mathematically like this: $0 \leqslant P(E) \leqslant 1$.

So let this be a warning to you – if you are calculating the probability of an event, and your result is either negative, or greater than one, then you have made a mistake somewhere. Examiners have frequently met solutions to probability problems in the form of (say) $P(E) = 1.5$. Now not only has the candidate obviously performed the calculations incorrectly, but also demonstrated that he/she does not know that $P(E)$ cannot exceed one – you cannot be more certain than absolute certainty! So if you obtain an answer like this in an examination, and you cannot discover where you have gone wrong, please do state that your answer *is* wrong, and state *why* it is wrong!

Earlier, we stated that an experiment has N equally likely outcomes, n of which constitute event E. Hence, it must follow that $N - n$ of the outcomes would *not* constitute event E. We can now formulate:

$$P(E') \frac{N-n}{N} = 1 - \frac{n}{N}$$

So $P(E') = 1 - P(E)$

We have already discovered that the probability of drawing an ace from a pack of cards is $\frac{1}{13}$, so it must follow that the probability of not drawing an ace is $1 - \frac{1}{13} = \frac{12}{13}$. Later we will find this formula extremely useful.

1.4 The three approaches to probability

Well, we have now seen how to measure probability. However, we have so far been making an assumption without actually spelling it out. We have assumed that we not only know all the possible outcomes of an experiment, but also that we can weight the probability of each outcome in proportion to its likelihood. More importantly, we have assumed we can do both of these things *before the experiment is performed*. In other words, we assume a prior knowledge of the outcomes – we have been using the so-called *a priori* approach to probability. Although it is true that in many cases we will have the necessary information to use an a priori approach (it is true, for example, when considering games of chance), there are many cases in which such an approach cannot be used.

Suppose we have a large case of wood screws, and we wish to find the probability that one screw chosen at random is defective. Clearly, it is possible here to define all the events (the screw is either defective or it is not) but it is not possible to weight the events in proportion to their likelihood. The only way we can determine probability in this case is to draw a sample of N screws, and count the number of defectives (call this n); we can *estimate* the probability that a randomly chosen screw is defective as n/N. This is the so-called *empirical approach* – there is just no way of estimating the probability without drawing that sample!

An appreciation of these two approaches to probability helps to explain a problem that confuses so many students. The problem runs something like this: if I spin a penny 100 times, and on 99 occasions the coin shows heads, what is the probability that it will show heads on the next spin? Some people would argue that the outcome can be either a head or tail, and the coin has no memory of the 100 previous tosses. So the probability that the coin will show heads on the next spin must be $\frac{1}{2}$. Others would argue that the coin is more likely to show heads than tails, and would estimate the probability of obtaining a head on the next spin to be 99/100. Well, which approach is the correct one? Surprisingly, the answer is both! In the first case, we are using an a priori approach, reasoning that the experiment has produced a fluke result which does not detract from the fact that the coin is unbiased. In the second case, we are using an empirical approach, stating that the experimental evidence indicates that the coin is biased in favour of heads. Now ask yourself this: if you were a gambler, which approach would you prefer to use?

There is a third approach to probability that we must now examine. Suppose we wished to find the probability that a particular horse wins the Derby. Clearly we cannot use an a priori approach; nor can we use an empirical approach, as this would demand that the same race be repeated many times under identical conditions! The only way we can obtain this probability is to give personal 'gut feeling' of the horse's chances! This is the so-called *subjective* approach to probability, and it is the method used by bookmakers when fixing odds for a particular horse to win a race. Initially, the odds will be determined by the personal view of the bookmaker, and will be modified as the race approaches according to the collective, subjective views of the punters.

These three distinct approaches to probability raise an interesting philosophical problem. We know that $P(E)$ can never exceed one but does $P(E) = 1$ imply absolute certainty? It all depends on the approach used. If we use an a priori approach, then $P(E) = 1$ means E *must always* occur. However, using an empirical approach $P(E) = 1$ means that E *has always* occurred – which does not imply that it *must* occur in the future. Likewise, using a subjective approach, $P(E) = 1$ means that *we think that E will occur* – which again does not imply that it must occur.

Without doubt, the empirical and subjective approaches are more interesting and more useful than the a priori approach. However, in an introductory manual such as this, it is preferable to concentrate attention on a priori probability and, unless we state to the contrary, we assume an a priori approach is used.

1.5 The laws of probability

Let us suppose that we cast two dice, and add the scores of the dice. We could represent all the outcomes in a table like Fig. 1.1.

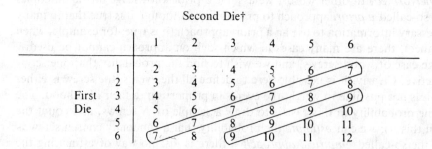

Second Die†

		1	2	3	4	5	6
	1	2	3	4	5	6	7
	2	3	4	5	6	7	8
First	3	4	5	6	7	8	9
Die	4	5	6	7	8	9	10
	5	6	7	8	9	10	11
	6	7	8	9	10	11	12

Figure 1.1

† Die is the singular, and dice the plural

The number in front of each row represents the possible scores of the first die, and the number at the head of each column represents the possible scores of the second die. The numbers in the main body of the table represent the sums of the two possible scores. So we see that if we score a total of 11, we have thrown either a six with the first die and a five with the second, or a five with the first die and a six with the second. We see, then, that there are 36 equally likely total scores ($N = 36$). Let us now define two events:

E_1 = the sum of scores is 7; E_2 = the sum of the scores is 9.

Now there are six ways of scoring a total of 7, so $P(E_1) = \frac{6}{36} = \frac{1}{6}$. Again there are four ways of scoring a total of 9, so $P(E_2) = \frac{4}{36} = \frac{1}{9}$. Now suppose we wished to find the probability that the sum of the scores is *either* 7 *or* 9. We can write the probability symbolically like this:

$$P(E_1 \cup E_2)$$

where $\cup$ is a shorthand way of writing 'either − or'. Consulting the table, we see that there are 10 ways of obtaining a total score of either 7 or 9, so $P(E_1 \cup E_2) = \frac{10}{36} = \frac{5}{18}$. Notice that the 10 ways are obtained by adding the number of ways for E_1 and E_2. This gives us our first law of probability: the so-called addition law:

$$P(E_1 \cup E_2) = P(E_1) + P(E_2)$$

a law which is true only if E_1 and E_2 are mutually exclusive. If the events are not mutually exclusive, then using this law will not yield the correct probability. Suppose we draw a card from a pack, and E_1 is that the card is an ace. So $P(E_1) = \frac{4}{52}$. If E_2 is that the card is a heart, then $P(E_2) = \frac{13}{52}$. If we want to find the probability that the card is either a heart or an ace, we notice that there are 52 equally likely outcomes, 16 of which would be either a heart or an ace (i.e. 13 hearts plus the three other aces). So the probability that the card is either a heart or an ace is $\frac{16}{52}$. Notice that if we applied the addition law, we would have obtained $\frac{13}{52} + \frac{4}{52} = \frac{17}{52}$, the wrong probability! So beware − before using this law make absolutely sure that the events are mutually exclusive!

Look again at the table we obtained earlier which refers to the sum of the possible scores from casting two dice. Notice that there are 36 equally likely total scores. We could deduce this as follows: there are six ways the first die can fall, each of which can combine with any one of the six ways that the second die can fall. So there are $6 \times 6 = 36$ ways that both dice can fall. Now suppose we have two piles of cards. The first pile contains the two red aces and the four kings, and the second pile contains the two red aces and the ace of spades and four kings. Suppose we draw a card from each pile − as there are six cards in the first pile and seven cards in the second, there will be $6 \times 7 = 42$ ways of drawing a pair of cards, one from each pile. We could represent the situation in a table like Fig. 1.2.

		Second Card						
		A♡	A♢	A♠	K♡	K♢	K♠	K♣
	A♡	1	2	3	4	5	6	7
	A♢	8	9	10	11	12	13	14
First	K♡	15	16	17	18	19	20	21
Card	K♢	22	23	24	25	26	27	28
	K♠	29	30	31	32	33	34	35
	K♣	36	37	38	39	40	41	42

Figure 1.2

In this table we have numbered all the possible 42 combinations of events 1,2,3, … etc., so combination 25, for example, means drawing the king of diamonds with the first card and the king of hearts with the second. Let us now define two events:

E_1 = the first card is an ace, so $P(E_1) = \frac{2}{6}$

E_2 = the second card is an ace, so $P(E_2) = \frac{3}{7}$

Suppose we wanted to find the probability that both cards were aces. We could write it symbolically:

$P(E_1 \cap E_2)$

where the symbol $\cap$ is a shorthand form for 'both ... and', notice that there are six ways of obtaining two aces, so $P(E_1 \cap E_2) = \frac{6}{42}$. Now we could have obtained this result by multiplying $P(E_1)$ and $P(E_2)$ together ($\frac{2}{6} \times \frac{3}{7} = \frac{6}{42}$). This gives us the second law of probability, the so-called multiplication law:

$P(E_1 \cap E_2) = P(E_1) \cdot P(E_2)$

a law which applies only if E_1 and E_2 are *independent* (i.e. as long as the outcomes in no way affect each other). We will discuss this point more fully later, but it is worth noting now that independent events cannot be mutually exclusive, and mutually exclusive events cannot be independent.

The second law of probability enables us to modify the first law to take account of events that are not mutually exclusive. A little earlier, we considered the case of drawing a card from a pack, calling E_1 that the event was an ace (so $P(E_1) = \frac{4}{52}$), and calling E_2 that the card is a heart (so $P(E_2) = \frac{13}{52}$). We stated that the probability that the card is either a heart or an ace $P(E_1 \cup E_2)$ is not $\frac{4}{52} + \frac{13}{52} = \frac{17}{52}$. Now why doesn't the addition law work? Surely the fault here is that the card we draw could be the ace of hearts *and we have counted this card twice* – once as a heart and once as an ace. So the probability that the card is either an ace or a heart is $\frac{17}{52}$ minus the probability that the card is the ace of hearts. Using the second law, the probability that the card is the ace of hearts is $P(E_1 \cap E_2) = \frac{4}{52} \times \frac{13}{52} = \frac{1}{52}$, so the probability that the card is either an ace or a heart is $\frac{17}{52} - \frac{1}{52} = \frac{16}{52}$, which agrees precisely with the result we obtained from first principles. We can now restate the addition law to take account of situations when the events are not mutually exclusive:

$P(E_1 \cup E_2) = P(E_1) + P(E_2) - P(E_1 \cap E_2)$

This is the so-called *general law of addition*, and it works whether the events are mutually exclusive or not (if the events are mutually exclusive, then they cannot both occur, so $P(E_1 \cap E_2)$ will be zero).

1.5.1 Venn diagrams

An alternative way of approaching probability problems is to use Venn diagrams. Firstly, we define a *sample space* as the set of all possible outcomes E. This is represented in the Venn diagram by a box. An event (which we have already defined as the outcome of an experiment) would be a *subset* of the sample space, and is represented by a circle. To take a simple example, let the sample space be represented by a pack of playing cards from which we intend to draw a card. Let us define two events:

E_1 = the card is a heart; E_2 = the card is an ace.

The Venn diagram of this experiment would look like Fig. 1.3. The diagonally shaded area represents the event: the card is both an ace and a heart (i.e. it is the ace of hearts). *If both events can occur, then the subsets must intersect.* The horizontally shaded area represents the event: the card is neither an ace nor a heart.

Now let us define N as the number of equi-likely outcomes comprising the sample space and $n(E)$

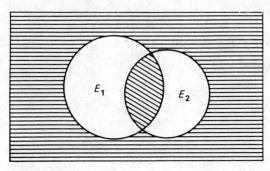

Figure 1.3

as the number of equi-likely outcomes comprising the event *E*. If these numbers are inserted on the Venn diagram, then it is possible to deduce probabilities for the events either singly or in combination.

Example 1

Suppose 1200 items are randomly selected from a production line, and examined for defects. Of the items, 60 had a dimension defect and 50 had a strength defect, while 20 had both defects. We require to know the probability that on the basis of this sample evidence a randomly chosen item (a) is defective, (b) is non-defective.

For this example,

$$N = 1200$$
$$n(\text{dimension}) = 60$$
$$n(\text{strength}) = 50$$
$$n(\text{dimension and strength}) = 20$$
$$\text{so, } n(\text{dimension only}) = 60 - 20 = 40$$
$$\text{and } n(\text{strength only}) = 50 - 20 = 30$$

The Venn diagram for the experiment would look like Fig. 1.4.

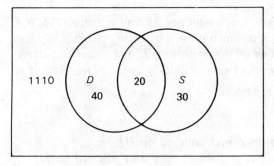

Figure 1.4

You should realise that

$$n(\text{not defective}) = 1200 - 20 - 40 - 30 = 1110.$$

We can now deduce that $p(\text{item defective}) = \dfrac{20 + 40 + 30}{1200} = 0.075$

and that $p(\text{item non-defective}) = \dfrac{1100}{1200} = 0.925$

1.5.2 Applications of the laws of probability

Example 2

Two types of metal A and B, which have been treated with a special coating of paint, have probabilities of $\frac{1}{4}$ and $\frac{1}{3}$ respectively of lasting four years without rusting.

If both types of metal are given the special coating on the same day, what is the probability that (a) both last 4 years without rusting; (b) at least one of them lasts 4 years without rusting.

For part (a), we can use the multiplication law to obtain a probability of $\frac{1}{4} \times \frac{1}{3} = \frac{1}{12}$ that both last for 4 years. Turning to the second part, we should first notice that four distinct outcomes are possible

(a) both last 4 years   (probability is $\frac{1}{4} \times \frac{1}{3} = \frac{1}{12}$)
(b) A lasts, B does not   (probability is $\frac{1}{4} \times \frac{2}{3} = \frac{2}{12}$)
(c) B lasts, A does not   (probability is $\frac{3}{4} \times \frac{1}{3} = \frac{3}{12}$)
(d) neither lasts   (probability is $\frac{3}{4} \times \frac{2}{3} = \frac{6}{12}$)

Now any of outcomes (a), (b) or (c) satisfies the condition that at least one lasts 4 years, so the probability we require is $\frac{1}{12} + \frac{2}{12} + \frac{3}{12} = \frac{6}{12} = \frac{1}{2}$. Notice that we could have calculated the probability more directly by using the fact that:

$P(\text{at least one lasts}) = 1 - P(\text{neither lasts})$

This second method can save quite a lot of time, and you should always use it in preference to writing out all the possible outcomes.

Example 3

An item is made in three stages. At the first stage, it is formed on one of four machines, A, B, C or D, with equal probability. At the second stage, it is trimmed on one of three machines, E, F or G, with equal probability. Finally, it is polished on one of two polishers, H and I, and is twice as likely to be polished on the former as this machine works twice as quickly as the other. Required:

(a) What is the probability that an item is:
    (*i*) polished on H?
    (*ii*) trimmed on either F or G?
    (*iii*) formed on either A or B, trimmed on F and polished on H?
    (*iv*) either formed on A and polished on I, or formed on B and polished on H?
    (*v*) either formed on A or trimmed on F?

(b) Suppose that items trimmed on E or F are susceptible to a particular defect. The defect rates on these machines are 10 and 20 per cent respectively. What is the probability that an item found to have this defect was trimmed on F? (ACA)

First, we shall determine the probabilities for each machine. As the formation stage is equally likely to occur on any one of the four machines, we have:

$$P(A) = P(B) = P(C) = P(D) = \tfrac{1}{4}$$

Again, trimming is equally likely to occur on any one of the three machines, so:

$$P(E) = P(F) = P(G) = \tfrac{1}{3}$$

Polishing is twice as likely to occur on machine H as machine I, so:

$$P(H) = \tfrac{2}{3}, \quad P(I) = \tfrac{1}{3}$$

(a) (i) The probability that an item is polished on $H = P(H) = \tfrac{2}{3}$.

(ii) The probability that an item is trimmed on either F or $G = P(F \cup G) = P(F) + P(G) = \tfrac{2}{3}$.

(iii) Formed on either A or B, trimmed on F and polished on H

$$= P(A \cup B) \cap F \cap H$$
$$= [P(A) + P(B)] \cdot P(F) \cdot P(H)$$
$$= (\tfrac{1}{4} + \tfrac{1}{4}) \times \tfrac{1}{3} \times \tfrac{2}{3} = \tfrac{1}{9}$$

(iv) Either formed on A and polished on I or formed on B and polished on H

$$= [(A \cap I) \cup (B \cap H)]$$
$$= [P(A) \cdot P(I)] + [P(B) \cdot P(H)]$$
$$= (\tfrac{1}{4} \times \tfrac{1}{3}) + (\tfrac{1}{4} \times \tfrac{2}{3})$$
$$= \tfrac{1}{4}$$

(v) Either formed on A or trimmed on F. These events are *not* mutually exclusive as it is possible to form on A *and* trim on F, so we need the general rule of addition, i.e.:

$$P(A \cup F) = P(A) + P(F) - P(A \cap F)$$
$$= \tfrac{1}{4} + \tfrac{1}{3} - (\tfrac{1}{4} \times \tfrac{1}{3})$$
$$= \tfrac{1}{2}$$

(b) Dealing with the second part of this question, we notice that F is twice as likely to produce a defective item as is machine E. So, given that the item is defective, the probability that it was trimmed on F must be $\tfrac{2}{3}$.

Example 4

In the past, two building contractors, A and B, have competed for twenty building contracts of which ten were awarded to A and six were awarded to B. The remaining four contracts were not awarded to either A or B. Three contracts for buildings of the kind in which they both specialise have been offered for tender.

Assuming that the market has not changed, find the probability that

(a) A will obtain all three contracts;
(b) B will obtain at least one contract;
(c) Two contracts will not be awarded to either A or B;

(d) A will be awarded the first contract, B the second, and A will be awarded the third contract.

(CIMA)

The probability that A gets the contract $P(A) = \frac{10}{20} = \frac{1}{2}$, the probability that B gets the contract is $P(B) = \frac{6}{20} = \frac{3}{10}$, and the probability that neither A nor B gets the contract is $P(A \cup B)' = \frac{4}{20} = \frac{1}{5}$.

(a) The probability that A will obtain all three contracts is $\frac{1}{2} \times \frac{1}{2} \times \frac{1}{2} = \frac{1}{8}$.

(b) The probability that B will obtain at least one contract

$$= 1 - P(B \text{ obtains no contracts})$$
$$= 1 - (\frac{7}{10} \times \frac{7}{10} \times \frac{7}{10}) = \frac{657}{1000}$$

(c) The probability that a contract is awarded to A or $B = P(A \cup B) = \frac{4}{5}$. We require to know the probability that two contracts will not be awarded to either A or B; this is equivalent to finding the probability that one of the contracts is awarded to either A or B. But the contract awarded to either A or B could be either the first, the second or third contract. So the probability we require is:

$\frac{4}{5} \times \frac{1}{5} \times \frac{1}{5}$ (A or B wins the first contract)

plus $\frac{1}{5} \times \frac{4}{5} \times \frac{1}{5}$ (A or B wins the second contract)

plus $\frac{1}{5} \times \frac{1}{5} \times \frac{4}{5}$ (A or B wins the third contract)

$$= 3 \times \frac{4}{5} \times \frac{1}{5} \times \frac{1}{5} = \frac{12}{125}$$

(d) The probability that A is awarded the first contract, B the second and A the third is $\frac{1}{2} \times \frac{3}{10} \times \frac{1}{2} = \frac{3}{40}$.

1.6 Tree diagrams

One of the major problems that occur when dealing with probability is that of ensuring that all logical possibilities are considered. In fact, this is probably the most common student error when dealing with this topic. A tree diagram is a device to help avoid such errors: it looks something like a tree, each branch of which represents one logical possibility. We shall illustrate tree diagrams by means of an example.

Buggsy Flynn owns 60 per cent of protection rackets and 80 per cent of illegal gambling in Chicago, and he is informed that the police are about to investigate both these activities. Police records show that 70 per cent of investigations into the protection racket and 90 per cent of investigations into gambling lead to court action. What is the probability that Buggsy will end up in court as a result of the investigations? (You may care to attempt to solve this problem before reading on.)

This problem involves an examination of whether Buggsy is investigated and whether he is charged. We can begin the tree diagram by calculating the probability of an investigation into Buggsy's protection rackets. Since he owns only 60 per cent of the protection in Chicago there is a 0.6 probability that the police investigation will concern him, and the position can be illustrated as in Fig. 1.5(a). Notice that the probabilities have been inserted on to the diagram. We can now add the possibility of an investigation into illegal gambling on to it as in Fig. 1.5(b).

We now have four routes on our tree, and if we multiply the probabilities along each route, then we have deduced the probability of certain events occurring at each point. For example, point 1 represents the situation where Buggsy is investigated on both activities and we find the probability of this happening is 0.48; point 3 is the situation where Buggsy is not investigated on his protection racket but is investigated on his illegal gambling, and the probability of this 0.32. Notice that the four points cover all the logical possibilities of investigations, so the probabilities must total 1 $(0.48 + 0.12 + 0.32 + 0.08 = 1.0)$.

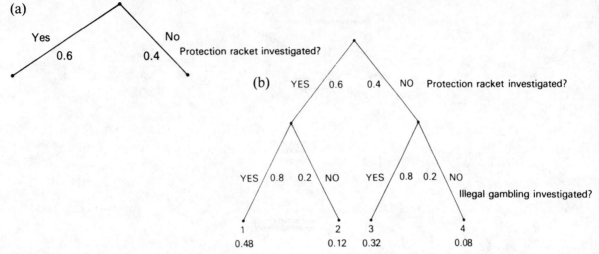

Figure 1.5

We will now insert on the the diagram the possibilities of court action resulting from a protection racket investigation (Fig. 1.6). Notice particularly points *a* and *b* on this tree diagram. Since on this branch the protection racket has not been investigated, there can be no court action on that charge and hence the probabilities at points *a* and *b* remain 0.32 and 0.08.

Finally, we can insert the possibility of court action on an illegal gambling charge and find the probability of each logical possibility (Fig. 1.7). Notice again points *a*, *b* and *c*. In each of these branches there has been no investigation of illegal gambling. Hence there can be no court charge and the probabilities remain as at the previous junction.

In Figure 1.7 we have underlined every logical possibility that results in Buggsy landing up in court. The probability of his facing a court charge is the sum of the probabilities of all of these, i.e.:

$$0.3024 + 0.1296 + 0.084 + 0.288 = 0.804$$

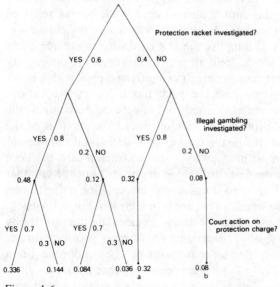

Figure 1.6

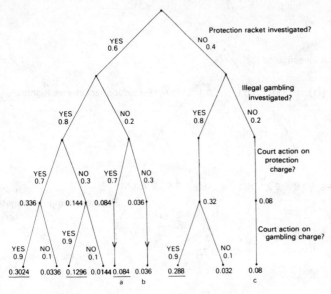

Figure 1.7

1.7 Conditional probability

In the first section, we stated that events cannot be both mutually exclusive and independent. This should not be taken to mean that if events are not mutually exclusive then they must be independent: there is a third category that we must now examine. When we state that events are independent, we mean that the outcome of one event in no way affects the outcome of the other. Now there are many cases where this is not true: the outcome of the second event is *conditional* on the outcome of the first event. Two examples may clarify this point.

Suppose we draw a card from a well-shuffled pack, and this card happens to be an ace. We now draw a second card – what is the probability that this card is also an ace? Well, having removed one card already, there must be 51 equally likely outcomes left, three of which would yield an ace. So the probability of an ace with the second card is $\frac{3}{51}$. Suppose the first card was not an ace: there would be 51 equally likely outcomes of which four would yield an ace. In this case, the probability of an ace with the second card would be $\frac{4}{51}$. In the first case, we are drawing from a pack with a lower proportion of aces than originally, and in the second case the pack has a higher proportion. Whichever way you consider this problem, the outcome of the first trial affects the outcome of the second. Now this raises an interesting philosophical problem: we cannot predict the outcome of the second event until we know the outcome of the first. If I deal a card to you, and you do not reveal it to me, the probability that I deal myself an ace must be $\frac{4}{52}$. Why is this so when a card has been removed from the pack? It is the *information* from the first card that is important, not the fact that it has been dealt. In this case, as you do not reveal the card it has a zero information value to me: from the information viewpoint, it is irrelevant to me whether the card is in your hand or in the pack. However, if your card is revealed to me, then I can use this information to calculate the probability that I deal myself an ace – the probability will be $\frac{4}{51}$ or $\frac{3}{51}$, depending on whether you have an ace or not. We need a new notation to take into account the fact that events can be conditional upon each other. If we have two events E_1 and E_2 then $P(E_2 \mid E_1)$ is the probability that E_2 occurs given that E_1 has occurred. So we can now modify our multiplication law to take account of conditional probability:

$$P(E_1 \cap E_2) = P(E_1) \cdot P(E_2 \mid E_1)$$

If E_1 is draw an ace with the first card, then $P(E_1) = \frac{1}{13}$. If E_2 is draw an ace with the second card, then $P(E_2 \mid E_1) = \frac{3}{51}$. The probability of drawing an ace with both cards is $P(E_1 \cap E_2) = \frac{1}{13} \times \frac{3}{51} = \frac{1}{221}$.

Now let us consider a second example. Suppose we have a box of ten machine parts, three of which are defective. From this we draw a sample of two parts – what is the probability they are both defective? If E_1 is that the first part is defective, and E_2 is the second part is defective, then the events are conditional. $P(E_1) = \frac{3}{10}$, $P(E_2 \mid E_1) = \frac{2}{9}$ and $P(E_1 \cap E_2) = \frac{3}{10} \times \frac{2}{9} = \frac{1}{15}$. Can you see that if we had replaced the first part before drawing the second then E_2 would not be conditional on E_1, and $P(E_2)$ would also be $\frac{3}{10}$? Using the statisticians' jargon, we would say that *sampling without replacement makes the events conditional*. Is this always true? We might have drawn two parts from a very large consignment indeed, and it would seem rather pedantic to state that the consignment is poorer in defectives if the first item drawn is defective. Moreover, the proportion of defectives in a very large consignment can only be an estimate. When sampling from a large population, then E_1 and E_2 can for all intents and purposes be considered independent.

The importance of conditional probability is that it enables us to modify our probability predictions in the light of any additional information that is made available.

Example 5

When exploration for oil occurs, a test hole is drilled. If as a result of this test drilling it seems likely that really large quantities of oil exist (a bonanza), then the well is said to have structure. Examination of past records reveals the following information:

probability (structure and bonanza)   0.20
probability (structure but no bonanza)   0.15
probability (no structure but a bonanza)   0.05
probability (no structure and no bonanza)   0.60

We can put this information into a table like this:

	Structure	No structure	
Bonanza	0.20	0.05	0.25
No bonanza	0.15	0.60	0.75
	0.35	0.65	1.00

Calculating the row totals and the column totals (often called the marginal probabilities), we can deduce that

probability (bonanza) = 0.25

since a bonanza can occur either with or without structure. Similarly,

probability (no bonanza) = 0.75
probability (structure)   = 0.35
probability (no structure) = 0.65

Such probabilities are known as prior probabilities – they are obtained from past records. But suppose we are given some additional information – in particular, suppose we know that a well has been sunk and structure is revealed. The number of possible outcomes has been reduced to two. We can ignore the 'no structure' column of the table and conclude that the weighting for a 'bonanza' is 0.2 and the weighting for 'no bonanza' is 0.15. So probability (bonanza | structure)

$$= \frac{0.2}{0.2 + 0.15} = 0.571$$

and probability (no bonanza | structure)

$$= \frac{0.15}{0.2 + 0.15} = 0.429$$

Now suppose we were given the information that the hole does not have structure. We can deduce that:

probability (bonanza | no structure) $\qquad = \dfrac{0.05}{0.65} = 0.077$

and probability (no bonanza | no structure) $= \dfrac{0.6}{0.65} = 0.923$

The probabilities we have just calculated are called posterior probabilities – they cannot be established until after something has happened (i.e. the test hole drilled). The knowledge whether the test hole has structure or not is valuable. Without this knowledge, we would estimate the probability of obtaining a bonanza at 0.25, but once the test hole is drilled we can revise our estimates. If the test hole reveals structure, then the probability of a bonanza rises to 0.571; if it does not reveal structure, the probability of a bonanza falls to 0.077.

Sometimes it is more useful to use tree diagrams to solve conditional probability problems.

Example 6

Suppose we have 100 urns. Type 1 urn (of which there are 70) each contains 5 black and 5 white balls. Type 2 urn (of which there are 30) each contains 8 black and 2 white balls. An urn is randomly selected and a ball is drawn from that urn. If the ball chosen was black, what is the probability that the ball came from a type 1 urn?

First, we notice that:

P(urn type 1)   = 0.7
P(urn type 2)   = 0.3
P(black | type 1) = 0.5
P(white | type 1) = 0.5
P(black | type 2) = 0.8
P(white | type 2) = 0.2

We can now draw the tree diagram (Fig. 1.8) and find the prior probabilities. Now given that a black ball was drawn, the logical possibilities of the tree are reduced from four to the two arrowed. So the probability that the ball came from a type 1 urn is:

$$\frac{0.35}{0.35 + 0.24} = 0.593$$

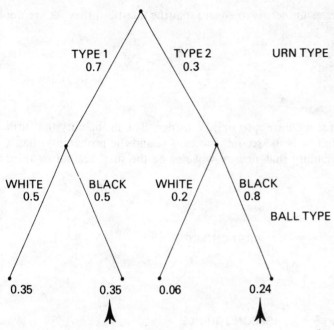

Figure 1.8

1.7.1 Bayes' theorem

An alternative approach to using tree diagrams to solve problems involving conditional probability is to use Bayes' Theorem. Suppose that a sample space consists of a number of mutually exclusive events A_1, A_2..., A_n. Suppose that B is any other event, then if we wish to calculate the probability that any of the events A_i occurs given that the event B has occurred, we can use the expression

$$P(A_i \mid B) = \frac{P(A_i)P(B \mid A_i)}{P(A_1)P(B \mid A_1) + P(A_2)P(B \mid A_2) + \cdots + P(A_n)P(B \mid A_n)}$$

In the previous example, let A_1 be the probability that urn type 1 is chosen and A_2 be the probability that urn type 2 is chosen. Hence

$P(A_1) = 0.7$
$P(A_2) = 0.3$

If we say that the event B represents the event that a black ball was chosen (i.e. the event we were given has occurred), then

$P(B \mid A_1) = 0.5$
$P(B \mid A_2) = 0.8$

Substituting these values in the above equation

$$P(A \mid B) = \frac{0.7 \times 0.5}{0.7 \times 0.5 + 0.3 \times 0.8}$$
$$= 0.593$$

When solving problems involving conditional probability, you should use either tree diagrams or Bayes' Theorem, whichever comes the easier to you.

One of the great difficulties for statistics examiners is to ensure that the questions they set are not ambiguous. Consider this case.

Example 7

Firm A is one of many firms competing for a contract to build a bridge. The probability that firm A is the first choice is $\frac{1}{9}$; the probability that it is the second choice is $\frac{1}{3}$, and the probability that it is the third choice is $\frac{1}{2}$. What is the probability that firm A will not be the first, second or third choice?

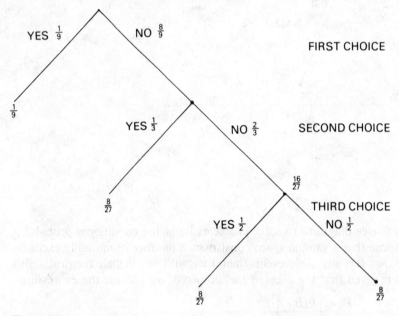

Figure 1.9

The ambiguity in this question arises from the precise meaning of the expression 'second choice' and 'third choice'. Are these probabilities joint probabilities, or are they conditional? Suppose we say that the probabilities are conditional – then the $\frac{1}{3}$ probably means that firm A is the second choice, given that it is not the first choice. The question can then be approached as in the tree diagram in Fig. 1.9.

So, assuming the probabilities are conditional, the probability that the firm is neither the first, second nor third choice is $\frac{8}{27}$.

On the other hand, if the probabilities are joint probabilities, the probability of $\frac{1}{3}$ is the probability that the firm is not the first choice but is the second. Think carefully about this distinction. If the probabilities are conditional we say the the probability that the firm is the second choice given that it is not the first is $P(\text{not the first}) \times P(\text{it is the second}) = \frac{8}{9} \times \frac{1}{3} = \frac{8}{27}$. But if the probabilities are joint, we are given the probability that the firm is not the first choice and the probability that it is (both not the first and is the second). That is, we now say $\frac{8}{9} \times P(\text{it is the second}) = \frac{1}{3}$. The question would be approached like this.

We will calculate the conditional probability that firm A is the second choice, given that it is not the first, as follows:

P(it is not the first) $\times$ P(it is the second | not the first) = P(it is both not the first and is the second)

that is:

$\frac{8}{9} \times$ conditional probability $= \frac{1}{3}$

conditional probability $= \frac{1}{3} \times \frac{9}{8} = \frac{3}{8}$

Thus we can build up our tree diagram as in Fig. 1.10.

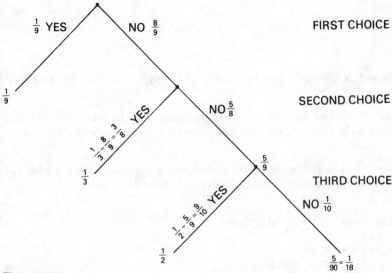

$\frac{1}{9}$ YES   NO $\frac{8}{9}$ FIRST CHOICE

$\frac{1}{9}$

$\frac{1}{3} \times \frac{8}{9} = \frac{3}{8}$ YES   NO $\frac{5}{8}$ SECOND CHOICE

$\frac{1}{3}$

$\frac{5}{9}$

$\frac{1}{2} \times \frac{5}{9} = \frac{9}{10}$ YES   THIRD CHOICE

NO $\frac{1}{10}$

$\frac{1}{2}$ $\frac{5}{90} = \frac{1}{18}$

Figure 1.10

We could deduce the probability that the firm is not the first, second nor third choice by subtraction:

$1 - (\frac{1}{9} + \frac{1}{3} + \frac{1}{2}) = \frac{1}{18}$

but it is more informative to work through conditional probabilities that we can calculate. Thus the probability that the firm is not the first choice is $\frac{8}{9}$. So:

P(firm A is second choice | not first choice) $= \frac{1}{3} \div \frac{8}{9} = \frac{3}{8}$
P(firm A is not second choice | not first choice) $= 1 - \frac{3}{8} = \frac{5}{8}$
P(firm A is neither first nor second choice) $= \frac{8}{9} \times \frac{5}{8} = \frac{5}{9}$
P(firm A is third choice | not first or second choice) $= \frac{1}{2} \div \frac{5}{9} = \frac{9}{10}$
P(A is not third choice | neither first nor second) $= 1 - \frac{9}{10} = \frac{1}{10}$
P(firm A is not first nor second nor third choice) $= \frac{1}{10} \times \frac{5}{9} = \frac{1}{18}$
the same answer as before

Now a few words to conclude this chapter. The concepts underlying probability are extremely easy to understand, and there are very few rules that must be learnt. However, putting these rules (or combinations of these rules) into practice can be very tricky indeed. You would be well advised to re-read the section on the laws of probability to make absolutely sure that you understand them. Also, make certain that you understand the logic employed in the previous examples, before you attempt the following Review questions. Before you attempt any calculations on them, you should be certain you know the way or ways the desired event can occur (for example, 'at least one' means one or more). Also, you should decide on the relationship between the events – whether they are mutually exclusive, independent, or conditional. Suppose that, after all this, you still cannot cope with probability – what then? Well, at least you can console yourself with the knowledge that although some students find probability a highly stimulating intellectual exercise, by far the bulk of them find probability problems the nearest thing to purgatory! Frequently you will find that although you can see that your answer is wrong, it is extremely difficult to see just where you have gone wrong! Fortunately, a lack of capability at solving probability problems need not hold you up in a statistics course, as a knowledge of what probability means and how it is measured are the important things.

Review questions

1.1 Two independent events A and B are such that $P(A) = 0.2$ and $P(B) = 0.4$. Find (a) $P(A \cap B)$, (b) $P(A \cup B)$.

1.2 An amplifier circuit is made up of three transistors. The probabilities that the three transistors are defective are $\frac{1}{20}$, $\frac{1}{25}$ and $\frac{1}{50}$ respectively. Calculate the probability that (a) the amplifier works, (b) that the amplifier has one defective transistor.

1.3 The probability that machine A will be performing a useful function in five years' time is $\frac{1}{4}$ while the probability that machine B will still be operating usefully at the end of the same period is $\frac{1}{3}$.
   Find the probability that in five years' time:

   (a) both machines will be performing a useful function;
   (b) neither will be operating;
   (c) only machine B will be operating;
   (d) at least one of the machines will be operating   CIMA

1.4 A businessman estimates that the probability of gaining an important contract to build a factory is 0.65 If this contract is obtained he will certainly have a probability of gaining a further contract of building an associated computer block – he estimates that this probability is 0.8. If he fails to obtain the contract to build the factory, he will not be asked to deal with the computer block construction. As an alternative to the factory/block contract, there is a probability of 0.35 that the businessman could obtain the contract to build an office block – he could only deal with this if the first set of contracts was not obtained. What is the probability that both factory and computer block contracts will be gained? Also, what is the probability that the businessman will obtain either the computer block contract or the office block contract?

1.5 The table given below shows a frequency distribution of the lifetimes of 500 light bulbs made and tested by ABC Limited:

Lifetime (hours)	Number of light bulbs
400 and less than 500	10
500 and less than 600	16
600 and less than 700	38
700 and less than 800	56
800 and less than 900	63
900 and less than 1000	67
1000 and less than 1100	92
1100 and less than 1200	68
1200 and less than 1300	57
1300 and less than 1400	33
	500

   (a) Using the information in this table, construct an ogive.

(b) Determine the percentage of light bulbs whose lifetimes are at least 700 hours but less than 1,200 hours.

(c) What risk is ABC Limited taking if it guarantees to replace any light bulb which lasts less than 1,000 hours?

(d) Instead of guaranteeing the life of the light bulb for 1,000 hours, ABC Limited suggests introducing a 100-day money-back guarantee. What is the probability that refunds will be made, assuming the light bulb is in use:

(i) 7 hours per day;

(ii) 11 hours per day?

1.6 A subassembly consists of three components A,B,C. Tests have shown that failures of the subassembly were caused by faults in one, two and sometimes all three components. Analysis of 100 subassembly failures showed that there were 70 faulty components A, 50 faulty components B, and 30 faulty components C. Of the failures 44 were caused by faults in two components only (i.e. A and B, or A and C, or B and C) and 10 of the 44 faults were in B and C.

If a faulty component is randomly selected, find the probability that it has faults in:

(a) all three components;

(b) component A on its own.

1.7 In a particular factory, an automatic process identifies defective items produced. Defectives can be classified as lacking strength, incorrect weight or incorrect diameter. A random sample 1,000 items were checked and the following results recorded:

120 have a strength defect;
80 have a weight defect;
60 have a diameter defect;
22 have strength and weight defects;
16 have strength and diameter defects;
20 have weight and diameter defects;
8 have all three defects.

Find the probability that a randomly chosen item

(a) is not defective;

(b) has exactly two defects.

1.8 An analysis of the national origins of the employees of a manufacturing company produced the following results:

English	488
Scots	230
Irish	112
Welsh	108
American	30
Other European	16

If two employees of this company are chosen at random, state the probability that

(a) both will be other European;

(b) both will be English;

(c) both will be either Irish or Welsh;

(d) at least one of the two will be a Scot.

1.9 The following results were obtained from interviews with 500 people who changed their cars recently, cars being classified as large, medium or small in size, depending on their length.

| | Present car size | | | |
Previous car size	Large (>15′)	Medium (13′−15′)	Small (<13′)	Total
Large	75	47	22	144
Medium	36	75	69	180
Small	11	63	102	176
Total	122	185	193	500

Figure 1.11

Required:

(a) What is the probability that a randomly chosen person from the survey changed to
  (i) a smaller car?
  (ii) a larger car?

(b) What effect would such car-changing habits have generally on the size of the car owned in the future?

(c) What is the probability that a person from the survey selected at random who bought a large car, previously had a small or medium car?

1.10

| | Table: Number of boxes Defective electron tubes per box of 100 units | | | |
Firm	0	1	2	3 or more
Supplier A	500	200	200	100
Supplier B	320	160	80	40
Supplier C	600	100	50	50

From the data given in the above table, calculate the conditional probabilities for the following questions:

(a) If one box had been selected at random from this universe, what are the probabilities that the box would have come from Supplier A; from Supplier B; from Supplier C?

(b) If a box had been selected at random, what is the probability that it would contain two defective tubes?

(c) If a box had been selected at random, what is the probability that it would have no defectives and would have come from Supplier A?

(d) Given that a box selected at random came from Supplier B, what is the probability that it contained one or two defective tubes?

(e) If a box came from Supplier A, what is the probability that the box would have two or less defectives?

(f) It is known that a box selected at random has two defective tubes. What is the probability that it came from Supplier A; from Supplier B; from Supplier C?   ACCA

1.11 A bag contains 19 balls, each of which is painted in two colours. Four are red and white, seven are white and black, and eight are black and red. A ball is chosen at random and seen to be partly white. What is the probability that its other colour is red?

1.12 Look again at Example 4 in the text. Suppose E_1 is that oil is found in commercial quantities and E_2 is that the well has structure. Given that $P(E_1 \cap E_2) = 0.42$, $P(E_1 \cap E_2') = 0.18$, $P(E_1' \cap E_2) = 0.28$ and $P(E_1' \cap E_2') = 0.12$, what would you conclude?

1.13 A public company holds two accounts: an account A with a government department and an account B with a merchant bank. It has been established that on any given day there is a finite and distinct probability that each account will exceed its overdraft facilities. It may be assumed that the probability that either or both accounts exceed their overdraft facilities is 0.55 whilst the probability that account B exceeds its overdraft facility is 0.25.

(a) If the fluctuations of the accounts are independent, what is the probability that account A exceeds its overdraft facilities on any given day?

(b) Had the fluctuations of the accounts been dependent, what would have been the probability that account A exceeded its overdraft on any day in which account B also exceeded its overdraft facilities if the probability that both accounts exceeded their overdraft facilities had been 0.12?

1.14 Tom, Dick and Harry are candidates for the post of works manager. The managing director will recommend a candidate and the recommendation must be ratified by the board of directors. The probability that the managing director recommends Harry is 60 per cent, with 25 per cent for Dick and 15 per cent for Tom. The probabilities that the board ratifies are 40 per cent for Harry, 35 per cent for Tom, and 25 per cent for Dick.

(a) Find the probability that none of the candidates are appointed.

(b) Find for each candidate the probability that he is appointed given that the board has ratified the managing director's recommendation.

Note: Use a tree diagram to solve this problem.
You should also attempt to solve part (b) of this problem using Bayes' Theorem.

2 Probability distributions

2.1 Introduction

In Chapter 1, we examined the basic concepts of probability. We can now use our knowledge to examine aspects of probability under certain rigidly defined conditions. The main difficulty that you have probably experienced so far in dealing with probability problems is that there is no one, assured method of dealing with them. However, when the conditions underlying a problem match the conditions we shall examine in this chapter, there will be a well-defined route to the solution. So although the concepts explained here are probably more involved than in Chapter 1 you can rest assured that the applications of the concepts will be much simpler. The first task will be to examine what is meant by a probability distribution.

2.2 What is a probability distribution?

Suppose in an experiment we define all the outcomes, and choose one of them as an attribute. We could then form a frequency distribution by finding the probability that the attribute does not occur, occurs once, occurs twice – and so on. The distribution we have formed is called a *probability distribution*. An example will clarify what we mean by this. Suppose we spin a coin, and choose as our attribute the number of heads occurring. The probability distribution would look like this:

Number of Heads	Probability (frequency)
0	$\frac{1}{2}$
1	$\frac{1}{2}$
	$\overline{1}$

Notice that the sum of the frequencies is one, a feature of all probability distributions. This is because the number of times the attribute can occur (in this case zero or one) form mutually exclusive, collectively exhaustive events.

In forming the above probability distribution, we have been using *a priori* concepts: we calculated the probabilities *before* the experiment was performed. However, it is certainly possible to form a probability distribution using empirical evidence, i.e. after the experiment is performed. Suppose, for example, we know that in a large batch of components some of them will be defective. We decide to draw a sample of six items, so the sample could be free from defectives, or contain 1, 2, 3, 4, 5 or 6 defectives. We wish to find the probability of each outcome. Well, we could draw, say, 100 samples and count the number of defectives in each. Our result may look like this:

Number of defectives	0	1	2	3	4	5	6
Frequency	75	15	7	2	1	0	0

To form the probability distribution, we divide each of the frequencies by 100 (the total frequency). So we have:

Number of defectives	0	1	2	3	4
Probability	0.75	0.15	0.07	0.02	0.01

In this chapter we will examine certain standard types of probability distributions and use them as a basis for calculating a priori probabilities. You may feel that we are devoting too much time to a priori probability and ignoring the other forms. However, if we *suspect* that the outcome of an experiment would conform to one of the standard types of probability distributions, then we can make some very useful predictions. We can always check our assumptions by obtaining empirical probabilities and comparing them with a priori probabilities.

2.3   The binomial distribution

The first probability distribution we will examine is the so-called *binomial distribution*, and we shall examine how it arises by considering a very simple example. Let us suppose that three coins are tossed together and take as our attribute the number of heads occurring. Well, the probability that none of the coins is a head can be calculated very simply using the multiplication law:

$$P_{(0)} = \tfrac{1}{2} \times \tfrac{1}{2} \times \tfrac{1}{2} = \tfrac{1}{8}$$

Now let us calculate the probability that one of the coins shows a head. The problem here is that whereas there is only one event where no heads occur, there are a number of events where just one head occurs, namely HTT, THT, TTH (where H is a head occurring and T is a tail occurring). Fortunately, each of the events is equally likely ($\tfrac{1}{2} \times \tfrac{1}{2} \times \tfrac{1}{2}$), so:

$$P_{(1)} = 3 \times \tfrac{1}{2} \times \tfrac{1}{2} \times \tfrac{1}{2} = \tfrac{3}{8}$$

Two heads can occur in three equally likely ways (HHT, HTH, THH), so:

$$P_{(2)} = 3 \times \tfrac{1}{2} \times \tfrac{1}{2} \times \tfrac{1}{2} = \tfrac{3}{8}$$

but three heads can occur in one way only (HHH), so:

$$P_{(3)} = \tfrac{1}{2} \times \tfrac{1}{2} \times \tfrac{1}{2} = \tfrac{1}{8}$$

We have now calculated the full probability distribution, and we could represent it like this:

No. of heads occurring	Probability	*or* Probability
0	$\tfrac{1}{8}$	0.125
1	$\tfrac{3}{8}$	0.375
2	$\tfrac{3}{8}$	0.375
3	$\tfrac{1}{8}$	0.125
	1	1.000

A binomial distribution is concerned with two terms (hence its name) – the probability that the event we are considering occurs (we will call this p) and the probability that the event does not occur (we call this $1 - p = q$). We have just considered an example where $p = q$ (the probability of obtaining a head is identical to the probability of obtaining a tail), but we can use a very similar analysis for cases where p and q are unequal *as long as the experiment is performed three times*. Suppose we have a very large consignment of components, and we know from past experience that 10 per cent of them are defective. We randomly select three components from the consignment – what is the probability distribution of the number of defectives in the sample? We let p be the probability that an item in the sample is defective. Hence we have $q = 0.9$, $p = 0.1$.

$$
\begin{aligned}
P_{(0)} &= q \times q \times q = (0.9)^3 & &= 0.729 \\
P_{(1)} &= 3 \times q \times q \times p = 3q^2 p = 3 \times 0.9^2 \times 0.1 & &= 0.243 \\
P_{(2)} &= 3 \times q \times p \times p = 3qp^2 = 3 \times 0.9 \times 0.1^2 & &= 0.027 \\
P_{(3)} &= p \times p \times p = p^3 = 0.1^3 & &= \underline{0.001} \\
& & & \underline{1.000}
\end{aligned}
$$

Notice that we are using the results of the previous example. There is only one way for the sample to contain no defectives, and only one way for it to contain three defectives. There are three equally likely ways that the sample can contain one defective, and three equally likely ways that the sample can contain two defectives.

Examining the above examples, we can see that to calculate the individual probabilities in a binomial distribution, we must find the number of equally likely events comprising the outcome we require, and multiply this by the probability of any one of the events. Suppose we increased the sample size in the question above to five items, and wished to find the probability that three of them were defective. Now we know that this outcome can occur in a number of equally likely ways, and it is easy to calculate the probability of any one of these ways. For example,

$$ q \times q \times p \times p \times p = q^2 p^3 = (0.9)^2 (0.1)^3 = 0.00081 $$

is the probability that the first two items are non-defective and the last three are defective. We now need to know the number of equally likely ways of obtaining this outcome. We can write out all the events (as we did previously) but this would be very tedious – there are 10 equally likely events in this case. Also, we could never be sure that we have included all of the outcomes! How do you think we could go on writing out all the outcomes of drawing, say, two defectives if the sample contained 100 items?

2.3.1 Pascal's triangle

A very useful device for finding all the equally likely events in a binomial distribution is *Pascal's triangle*, and an example of it is constructed below.

							n = no. of items
1	1						
1	2	1					2 in the
1	3	3	1				3 sample
1	4	6	4	1			4
1	5	10	10	5	1		5
1	6	15	20	15	6	1	6

| Number of defectives | 0 | 1 | 2 | 3 | 4 | 5 | 6 |

Reading off from the triangle, we see that if we draw a sample of 5 items ($n = 5$) there is one way that a sample can contain no defectives, there are five equally likely ways that the sample can have one defective, 10 ways for it to have two defectives – and so on. Using this information, we can calculate the complete binomial distribution.

No. of defectives	No. of equally likely events	Probability of one of these events	Required probability	
0	1	$(0.9)^5$	$(0.9)^5$	$= 0.59049$
1	5	$(0.9)^4(0.1)$	$5(0.9)^4(0.1)$	$= 0.32805$
2	10	$(0.9)^3(0.1)^2$	$10(0.9)^3(0.1)^2$	$= 0.0729$
3	10	$(0.9)^2(0.1)^3$	$10(0.9)^2(0.1)^3$	$= 0.0081$
4	5	$(0.9)\ (0.1)^4$	$5(0.9)\ (0.1)^4$	$= 0.00045$
5	1	$(0.1)^5$	$(0.1)^5$	$= 0.00001$
				1.00000

So you can see that Pascal's triangle is a useful way of finding the number of equally likely events. But if you are going to use Pascal's triangle, it will be necessary for you to be able to construct it. The first thing you should notice is that the first and last number in any row is one. The rest of the numbers are obtained by adding the numbers in the previous row together in pairs. So the numbers for row six would be:

1	1 + 5	5 + 10	10 + 10	10 + 5	5 + 1	1
= 1	6	15	20	15	6	1

and the numbers in row seven would be:

1	1 + 6	6 + 15	15 + 20	20 + 15	15 + 6	6 + 1	1
= 1	7	21	35	35	21	7	1

The numbers that we are obtaining in the rows of Pascal's triangle are called the *binomial coefficients*, and certainly it is easy to obtain them. But suppose we are drawing a sample of 20 items. If we do not have a copy of Pascal's triangle available, it will be necessary to construct its first 20 rows, and this is going to be quite a task! It is possible, of course, but let us examine an alternative to Pascal's triangle which we think that, in the long run, you will find easier.

2.3.2 Combinations

Most gamblers are aware that if you want the bookmaker to choose a stated number of results (say three) from a larger number of selections (say six), he will do so. In fact, this is one of the most popular forms of betting both on the football pools and on the racetrack. So popular is it that many bookmakers issue tables telling how much it will cost to bet on any number of results chosen from a larger number. Such a table is a copy of Pascal's triangle, but instead of reading off 'the number of defectives' we read off 'number of correct results'. So gamblers can now look at the table and see that if they want to select six football teams and win the bet if any three of them draw, then they must make 20 bets. Bookmakers call these tables 'permutation tables', and instruct us to write our bets in the form 'Perm any 3 results from 6 selections = 20 bets'.

Now strictly, the bookmaker is wrong – when betting on three results from six we are dealing not with a permutation but with a combination, and we will now show you how to calculate how many combinations of r results there are if you make n selections. We can write this problem symbolically like this:

$$^nC_r$$

and to calculate the number of combinations we apply the formula:

$$^nC_r = \frac{n(n-1)(n-2)\ldots[(n-r)+1]}{r(r-1)(r-2)\cdots \times 1}$$

This formula looks very fearsome, but in fact it is very easy to apply. If we examine Pascal's triangle we see that if we select seven football teams to win, then there are 35 equally likely ways of just three of them winning. Now let us check this by using the formula for combinations. Here we have $n = 7$, $r = 3$, so $(n - r) + 1 = (7 - 3) + 1 = 5$, and

$$^7C_3 = \frac{7 \times 6 \times 5}{3 \times 2 \times 1} = 35$$

We are now in a position to derive a general expression for a binomial distribution. Suppose that within a population, a proportion p has a certain attribute (call it 'defective') — it follows that $1 - p = q$ is the proportion of non-defectives. Now suppose we draw a sample of n items.

The probability that the sample is free from defective items $= P_{(0)} = q^n$. The probability that the sample has one defective item is $P_{(1)} = {}^nC_1 q^{n-1} p$ (if one item is defective then $n - 1$ must be non-defective). The probability that the sample has two defective items is $P_{(2)} = {}^nC_2 q^{n-2} p^2$.

We could continue in this way until we reach the probability that all the items in the sample are defective, which must be $P_{(n)} = p^n$. Now we know that all the probabilities must add up to one, so:

$$q^n + {}^nC_1 q^{n-1} p + {}^nC_2 q^{n-2} p^2 + {}^nC_3 q^{n-3} p^3 + \cdots + p^n = 1$$

We also know that $q + p = 1$, so $(q + p)^n$ must equal one. Now we can write:

$$(q + p)^n = q^n + {}^nC_1 q^{n-1} p + {}^nC_2 q^{n-2} p^2 + {}^nC_3 q^{n-3} p^3 + \cdots + p^n$$

and call the right-hand side of this expression the *expansion of the binomial* $(q + p)^n$. The individual terms of the right-hand side give the probabilities of $0, 1, 2, \ldots, n$ defectives.

The great advantage of using the general expression for the binomial distribution is that we often do not wish to calculate the entire distribution, but just part of it. If we knew that a bag of seeds had a 95 per cent germination rate, and we planted 10 of these seeds, we could quite easily calculate the probability that (say) two fail to germinate. We have:

$$q = 0.95, \quad p = 0.05, \quad n = 10, \quad r = 2, \quad (n - r) + 1 = 9$$

$$P_{(2)} = \frac{10 \times 9}{1 \times 2} (0.95)^8 (0.05)^2 = 0.0746$$

We would like to point out that when we state that it is quite easy to calculate probabilities using the general expression for the binomial distribution, we really mean that it is easy to decide what to do. We are only too willing to admit that the actual arithmetic operations are most tedious.

2.3.3 The mean and standard deviation of a binomial distribution

Let us begin by calculating the probabilities of the binomial distribution $(\frac{3}{4} + \frac{1}{4})^5$ (for example, sampling in groups of five from a population with 25 per cent defectives). We shall use the general expression rather than Pascal's triangle.

$$(\tfrac{3}{4}+\tfrac{1}{4})^5 = (\tfrac{3}{4})^5 + {}^5C_1(\tfrac{3}{4})^4(\tfrac{1}{4}) + {}^5C_2(\tfrac{3}{4})^3(\tfrac{1}{4})^2 + {}^5C_3(\tfrac{3}{4})^2(\tfrac{1}{4})^3 + {}^5C_4(\tfrac{3}{4})(\tfrac{1}{4})^4 + (\tfrac{1}{4})^5$$

$$= (\tfrac{3}{4})^5 + 5(\tfrac{3}{4})^4(\tfrac{1}{4}) + \frac{5 \cdot 4}{2 \cdot 1}(\tfrac{3}{4})^3(\tfrac{1}{4})^2 + \frac{5 \cdot 4 \cdot 3}{3 \cdot 2 \cdot 1}(\tfrac{3}{4})^2(\tfrac{1}{4})^3 + \frac{5 \cdot 4 \cdot 3 \cdot 2}{4 \cdot 3 \cdot 2 \cdot 1}(\tfrac{3}{4})(\tfrac{1}{4})^4 + (\tfrac{1}{4})^5$$

$$= 0.2373 + 0.3955 + 0.2637 + 0.0879 + 0.0146 + 0.001$$

We can now calculate the mean and standard deviation of the number of defectives in samples of five items.

Number of defectives (x)	Probability (f)	fx	fx^2
0	0.2373	0	0
1	0.3955	0.3955	0.3955
2	0.2637	0.5274	1.0548
3	0.0879	0.2637	0.7911
4	0.0146	0.0584	0.2336
5	0.0010	0.0050	0.0250
	1.0000	1.2500	2.5000

$$\bar{x} = \frac{\Sigma fx}{\Sigma f} = \frac{1.25}{1} = 1.25 \text{ defectives per sample}$$

$$\sigma = \sqrt{\frac{\Sigma fx^2}{\Sigma f} - \left(\frac{\Sigma fx}{\Sigma f}\right)^2} = \sqrt{\frac{2.5}{1} - \left(\frac{1.25}{1}\right)^2}$$

$$= 0.968 \text{ defectives per sample}$$

In fact, there is no need to go through this procedure to obtain the mean and standard deviation. For any binomial distribution we know that:

$$\text{mean} = np$$
$$\text{standard deviation} = \sqrt{npq}$$

We have just been considering the distribution $(\tfrac{3}{4}+\tfrac{1}{4})^5$, in which $p=\tfrac{1}{4}$, $q=\tfrac{3}{4}$ and $n=5$. The mean is $np = 5 \times \tfrac{1}{4} = 1.25$, and the standard deviation is $\sqrt{5 \times \tfrac{1}{4} \times \tfrac{3}{4}} = 0.968$.

Before we leave the binomial distribution, one very important point must be emphasised (though it is hoped that many of you will have realised it already). A binomial distribution assumes that the events are independent, that is, when we are drawing the sample the outcome of the first item drawn in no way affects the outcome of the second item drawn. This is only true if we are sampling from a large population.

2.4 The Poisson distribution

Suppose we draw samples of two items from a population with 50 per cent of items defective, and samples of 100 from a population with 1 per cent of items defective. In both cases the mean is the same (one defective per sample) – but we could not expect to get the same probability distribution. In the first case, the number of defectives per sample can range between zero and two, but in the second case it can range between zero and 100. But if we draw large samples from a population containing a small proportion of defectives, then a very remarkable thing happens. The probability distributions for such samples tend to become the same *as long as the mean is constant*. To illustrate this feature we consider three cases: drawing samples of 100 from a population containing 1 per cent defectives; samples of 1,000 from a 0.1 per cent defective population; and samples of 10,000 from a 0.01 per cent defective population. In each case then the mean is the same (one defective per sample).

$$(0.99 + 0.01)^{100} = 0.366 + 0.370 + 0.185 + 0.061 + 0.014 + 0.003 + 0.001 + \cdots$$
$$(0.999 + 0.001)^{1,000} = 0.368 + 0.368 + 0.184 + 0.061 + 0.015 + 0.003 + 0.001 + \cdots$$
$$(0.9999 + 0.0001)^{10,000} = 0.368 + 0.368 + 0.184 + 0.061 + 0.015 + 0.003 + 0.001 + \cdots$$

It would seem reasonable to suppose that we could discover the probability distribution for such samples as long as we know the mean number of defectives per sample. A second probability distribution – the *Poisson distribution* – will enable us to do this.

Before we examine the Poisson distribution and discover how it works, let us examine another condition which calls for its use. So far, we have been considering a sample of known and determinable size, and counting the number of times that an event occurred. We could also count the number of times the event did not occur. However, there are many cases where we cannot count the number of times the event did not occur. Suppose, for example, we wished to investigate the incidence of industrial accidents in a particular trade. We could count the number of accidents within (say) one year and use this as an estimate of the mean number of accidents. But we cannot count the number of times the accident did not occur! For problems such as this, we would have to use the Poisson distribution, as we have no way of evaluating p, q or n.

The Poisson distribution looks like this:

$$e^{-x} \left[1 + x + \frac{x^2}{2!} + \frac{x^3}{3!} + \frac{x^4}{4!} + \cdots \right]$$

Let us look at this distribution and examine certain features of it that you might not have met before. First, notice the numbers 2!, 3!, 4! (pronounce them '2 factorial, 3 factorial', etc.). This is just a convenient way of writing 'multiply the number n by $(n-1)$, then by $(n-2)$ and so on until finally we multiply it by one'. So:

2! = 2 × 1
3! = 3 × 2 × 1
4! = 4 × 3 × 2 × 1
5! = 5 × 4 × 3 × 2 × 1

and so on. Secondly, there is the number e. This is a well-known constant and has a value of 2.7183 (to four decimal places). In fact, most books of mathematical tables will have a table giving values of e^{-x}. Thirdly, the expression contains x, which is the mean of the distribution. We obtain probabilities from the expression like this:

$$P_{(0)} = e^{-x} \times 1$$
$$P_{(1)} = e^{-x} \times x$$
$$P_{(2)} = e^{-x} \times \frac{x^2}{2!}$$
$$P_{(3)} = e^{-x} \times \frac{x^3}{3!}$$

and so on. Now as e is a constant, the only variable in the expression is the mean x. So provided we know the mean of a distribution, we should be able to calculate the probabilities. Let's try it and see.

In a particular industry, there are on average two fatal accidents per year. We want to find the probability that (a) the industry is free from fatal accidents and (b) the industry has three fatal accidents in a year.

In this case, we have $x = 2$, and the appropriate Poisson distribution is:

$$e^{-2}\left[1 + 2 + \frac{2^2}{2!} + \frac{2^3}{3!} + \cdots\right]$$

$$P_{(0)} = e^{-2} \times 1$$

$$P_{(3)} = e^{-2} \times \frac{2^3}{3!}$$

Consulting the tables at the back of the book, we see that $e^{-2} = 0.1353$. However, it is possible that you may be asked to calculate e^{-x} from first principles. To do this, we make use of the fact that

$$\log e^{-x} = \log 1 - x \log e$$

Using logarithm tables, $\log e = 0.4343$, so

$$\log e^{-2} = 0 - (2 \times 0.4343)$$
$$= \bar{1}.1314$$

and

$$e^{-2} = 0.1353$$

Returning to our example,

$$P_{(0)} = 0.1353$$

so the probability that the industry is free from fatal accidents is 0.1353, or 13.53 per cent.

$$P_{(3)} = 0.1353 \times \frac{2^3}{3!}$$

$$= 0.1804$$

The probability that the industry has three fatal accidents in a year is 18.04 per cent.

2.4.1 Conditions necessary for using the Poisson distribution

We can calculate probabilities using the Poisson distribution provided that we know the arithmetic mean of the distribution, and provided that p is small and n is large. A further condition is that the mean must remain constant. However, perhaps the most important condition is that the occurrence of the event must be purely at random. A good illustration of this is the flow of traffic along a highway. If we choose an isolated point on a highway and count the number of vehicles passing that point, then we would probably find a random flow of vehicles. However, if the point we choose is near a set of traffic lights, then we will not find a random flow − the flow will be 'bunched', i.e. it will be heavy when the lights show green. So we should be able to tell whether the occurrence of the event is random or not by using the Poisson distribution. In fact, this distribution was used for an important piece of statistical investigation during the Second World War. In 1944, London was subjected to bombardment by German V1 rockets (called doodlebugs). This was a pilotless vehicle packed with high explosive and equipped with sufficient fuel to carry it to London. When the fuel ran out, the vehicle would fall out of the sky and explode on impact. The problem was to decide whether the V1 rockets were guided to particular targets with a great degree of precision, or whether they fell on London at random. In other words, were they falling in clusters to a greater degree than could be ascribed to chance? R. D. Clark divided an area of 144 square kilometres into 576 equal squares, and counted the number of rockets falling in each square.

No. of bombs per square	0	1	2	3	4	5	Total
No. of squares	229	211	93	35	7	1	576

Now if we divide the frequencies by 576, we can find the empirical probability distribution of the number of bombs per square:

No. of bombs per square	0	1	2	3	4	5
Probability	0.3976	0.3663	0.1615	0.0608	0.0121	0.0017

If the bombs were falling at random, then we should be able to predict this probability distribution using the a priori Poisson distribution. To do this, we need to know the mean number of bombs per square.

x	f	fx	fx^2
0	0.3976	0.0000	0.0000
1	0.3663	0.3663	0.3663
2	0.1615	0.3230	0.6460
3	0.0608	0.1824	0.5472
4	0.0121	0.0484	0.1936
5	0.0017	0.0085	0.0425
	1.0000	0.9286	1.7956

For the moment, ignore the column headed fx^2. The mean is:

$$\frac{\Sigma fx}{\Sigma f} = \frac{0.9286}{1} = 0.93 \text{ bombs per square}$$

and the Poisson distribution we require is:

$$e^{-0.93}\left[1 + 0.93 + \frac{(0.93)^2}{2!} = \frac{(0.93)^3}{3!} + \frac{(0.93)^4}{4!} + \frac{(0.93)^5}{5!}\right]$$

Using the tables at the end of this manual, we find that:

$e^{-0.93} = 0.3946$ so

$P_{(0)} = 0.3946$

$P_{(1)} = 0.3946 \times 0.93 = 0.3670$

$P_{(2)} = 0.3946 \times \frac{(0.93)^2}{2!}$ or using the previous term

$P_{(2)} = 0.3670 \times \frac{0.93}{2} = 0.1706$

$P_{(3)} = 0.3946 \times \frac{(0.93)^3}{3!}$

$= 0.1706 \times \frac{0.93}{3} = 0.0529$

$$P_{(4)} = 0.3946 \times \frac{(0.93)^4}{4!}$$

$$= 0.0529 \times \frac{0.93}{4} = 0.0123$$

$$P_{(5)} = 0.3946 \times \frac{(0.93)^5}{5!}$$

$$= 0.0123 \times \frac{0.93}{5} = 0.0023$$

Let us now write the empirical and Poisson probabilities adjacent to each other so that we can compare them.

No. of bombs per square	0	1	2	3	4	5
Empirical probabilities	0.3967	0.3663	0.1615	0.0608	0.0121	0.0017
Poisson probabilities	0.3946	0.3670	0.1706	0.0529	0.0123	0.0023

We feel sure that you will agree that there is a fantastically good agreement between the empirical and Poisson probabilities, and we must conclude that the V1 rockets were falling at random over the area.

Before we leave the Poisson distribution, we should note that its mean equals its variance (do you remember that the variance is the square of the standard deviation?). Now as the empirical and Poisson probabilities show such close agreement, we would expect the mean of the empirical probabilities to be approximately equal to the variance. We have already obtained the mean – it is 0.9286. The variance is:

$$\frac{\Sigma fx^2}{\Sigma f} - \left(\frac{\Sigma fx}{\Sigma f}\right)^2$$

$$= \frac{1.7956}{1} - \left(\frac{0.9286}{1}\right)^2$$

$$= 0.9333$$

Again, we find a very close agreement between the mean and variance.

Of the two discrete probability distributions we have examined in this chapter, the Poisson distribution is the most widely used. It has been found to describe 'accidents', traffic flows, and the arrivals into queuing situations, very well indeed. Without doubt, both distributions are very important. However, the probability distribution that is the most important in statistical analysis is the one we will examine in the next chapter – the normal distribution.

Review questions

2.1 A large company's records show that they have an average of 6 per cent of their employees off work on any one day. They employ six van drivers. You may assume that the probability of absence from work of a van driver on any one day is the same as that for any other employee.

    (a) What is the probability that all their van drivers will be at work on a given day?
    (b) What is the probability that at least five of their van drivers will be at work on a given day?

2.2 If half the electorate are supposed to vote Tory, and 100 investigators each ask 10 people their voting intentions, how many will report three or less voting Tory in their sample?

2.3 Razor blades are sold in packets of five. The distribution below shows the number of faulty blades in 100 packets.

No. of faulty blades	0	1	2	3	4	5
No. of packets	84	10	3	2	1	0

Calculate the mean number of faulty blades per packet. Assuming the distribution is binomial, estimate the probability that a blade taken at random from a packet will be faulty. (Hint: mean = np.)

2.4 At a certain time of day, the number of telephone calls coming in to a particular switchboard follows a Poisson distribution with a mean of two calls per minute. Calculate the probability of more than four calls in a minute.

2.5 The number of failures per week, of a certain type of machine, has been found to follow a Poisson distribution with mean 0.5. What is the probability that a given machine has three or more failures in a given week?

A firm owns five of these machines. What is the distribution of the total number of failures per week?

2.6 The demand for a component is two per month, and forms a Poisson distribution. Stock is made up at the beginning of each month. What should be the stock level at the beginning of each month so that the probability of a stockout is less than 5 per cent? ($e^{-2} = 0.1353$.)

3　The normal distribution

3.1　Introduction

In the previous chapter, we were introduced to the concept of the probability distribution, and examined two such distributions – the binomial and Poisson distribution. With both of these distributions, the variable under consideration was *discrete* – i.e. could be a whole number only. We can, for example, use the binomial distribution to find the probability of obtaining five defectives in a random sample of 10 items: but we cannot use it to find the probability of obtaining 5.5 defectives! We will now turn our attention to a probability distribution that can handle continuous data. It can be used, for example, to find the probability that a randomly chosen man is taller than 6 feet or, what amounts to the same thing, the proportion of the male population taller than 6 feet. In order to do this we will need to know certain features of the population. In particular, we will need to know the population mean and standard deviation. We will begin this analysis by examining the way in which the mean and standard deviation can be used to compare two populations.

3.2　Standard scores

Acme Ltd is considering the profits that it made last year. Its Office Equipment Division earned a profit of £65,000 and its Domestic Appliance Division earned a profit of £55,000. At first sight, it it seems that the Office Equipment Division had the most successful year. Suppose, however, that we were told that the average profit earned in the office equipment industry was £50,000, and that the average profit earned in the domestic appliance industry was £45,000. Seemingly, the profit opportunities in office equipment are greater. How, then, can we compare the performance of the two divisions? If we were told that the standard deviation of profits in the office equipment industry was £10,000, then a profit of £65,000 is:

$$\frac{65,000-50,000}{10,000}$$

= 1.5 standard deviations above the mean. Suppose also that we know the standard deviation of profit earned in the domestic appliance industry was £5,000, then a profit of £55,000 is:

$$\frac{55,000-45,000}{5,000}$$

= 2 standard deviations above the mean. So we can see that in the domestic appliance industry the relative performance (i.e. the performance with respect to competitors) was better than in the office equipment industry. Data calculated in this way – in terms of standard deviations measured from the mean – are called *standard scores* or *Z scores*, and for any variate *x* drawn from a population with a mean μ and a standard deviation σ, we can calculate the *Z* score like this:

$$Z = \frac{x - \mu}{\sigma}$$

Notice that the *Z* score is in relative, not absolute, units; so they can be used to compare values in distributions that do not use the same unit of measurement.

Now *Z* scores are very important to statisticians: under certain conditions, they can be used to predict the proportion of a distribution that is more than, or less than, a certain measurement. Let us first examine the conditions necessary for us to be able to do this. The distribution must be continuous; it must be symmetrical and it must be bell-shaped. In other words, it must be shaped like the one in Fig. 3.1. We call such a distribution a *normal distribution*, and call its shape *mesokurtic*. Now you may (with some justification) think that these conditions are very stringent and will not often be met in practice. We will have more to say about this in later chapters, but for the moment we will concentrate on learning how to make predictions based on the normal distribution.

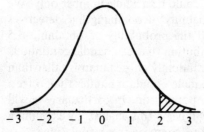

Figure 3.1

If you consult the tables at the end of this book you will find one headed the Normal Distribution Function. In the first column there is a list of *Z* scores ranging from 0.0 to 4.0. The second column gives the proportion of the normal distribution with a *Z* score greater than the corresponding figure in the first column. Suppose that we wanted to find the proportion in a normal distribution with a *Z* score greater than 2.0 (i.e. the proportion greater than two standard deviations above the mean). The shaded area in Fig. 3.1 represents the proportion diagrammatically. Finding 2.0 in the first column, we read that the corresponding figure in the second column is 0.02275 (or 2.275 per cent), so 2.275 per cent of normal distribution has a *Z* score greater than 2.0. This being so, it must follow that $100 - 2.275 = 97.725$ per cent of a normal distribution has a *Z* score less than 2.0 (we imply here that a negligible proportion will have a *Z* score *exactly* equal to 2.0).

The remaining columns in the table enable us to take an extra place of decimals when calculating the *Z* score. Suppose we wanted to find the proportion in a normal distribution with a *Z* score greater than 2.12. We find the required proportion at the junction of the row with a *Z* score 2.1, and the column headed 0.02, i.e. the proportion required is 0.01700. The normal distribution tables, then, follow the standard format followed by other mathematical tables.

Example 1

How can we apply this knowledge? Suppose we know that electric light bulbs have an average life of 2,000 hours and a standard deviation of 60 hours. Furthermore, we know lives of bulbs are

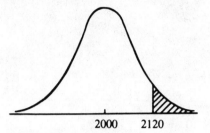

Figure 3.2

normally distributed (this would be a reasonable assumption) and we wish to know the proportion failing after 2,120 hours. We could represent the problem diagrammatically as in Fig.3.2

Notice that as the distribution is symmetrical, the mean bisects the distribution, and the mean, median and mode coincide. The shaded area will, of course, represent the proportion of bulbs failing after 2,120 hours, and only a small proportion will do so. The first thing we do is calculate the Z score:

$$Z = \frac{2,120-2,000}{60} = 2 \text{ standard deviations}$$

Now we already know that 2.275 per cent of items in a normal distribution have a Z score more than 2, and in this particular normal distribution any bulb with a life above 2,120 hours has a Z score more than 2. So it must follow that 2.275 per cent of bulbs will fail after 2,120 hours.

So far we have considered cases where the Z score is positive, but it is perfectly possible for a Z score to be negative. This will occur when an item under consideration has a value less than the arithmetic mean. Suppose, for example, we wished to find the proportion of bulbs with lives more than 1,850 hours. This involves finding the proportion of items with Z scores more than

$$\frac{1,850-2000}{60} = -2.5$$

If we consult the table, we notice that only positive Z scores are given – but the table can also be used for negative Z scores. A glance at Fig. 3.3(a) and 3.3(b) will confirm that as the distribution is symmetrical, the proportion of items with Z scores more than -2.5 is the same as the proportion with Z scores less than 2.5. Reading from the table, we see that 0.621 per cent of items have Z scores more than 2.5, so 99.379 per cent have Z scores less than 2.5, and so 99.379 per cent must have Z scores greater than -2.5. So we can see that 99.379 per cent of bulbs can be expected to fail after 1,850 hours.

Now you may find this all a bit confusing – sometimes the Z score is positive and sometimes it is negative; sometimes we read the percentage directly from the table, and sometimes we subtract it from 100. Can we offer some advice? When considering problems involving the normal

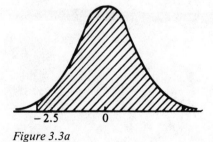

Figure 3.3a

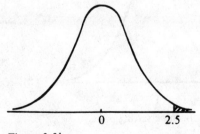

Figure 3.3b

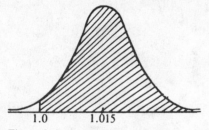

1.0   1.015

Figure 3.4

distribution, *always* draw a sketch of the problem and shade the area under consideration. You can then see at a glance whether the area required is greater or less than 50 per cent. If the area is less than 50 per cent, then the proportion in the table is the one required. If the area required is greater than 50 per cent subtract from 100 the percentage obtained from the table.

Example 2

Suppose we know that the weights of bags of flour are normally distributed with a standard deviation of 0.01 kilograms. The bags are marked 'weight not less than 1 kilogram', and the machine is set to fill the bags to an average weight of 1.015 kilograms. We require to know the proportion of bags that satisfy the claim printed on them. First, let us draw a sketch of the problem (see Fig. 3.4). The Z score is

$$\frac{1 - 1.015}{0.01} = -1.5$$

and looking up 1.5 in the table we obtain a proportion of 6.68 per cent. Now the diagram tells us immediately that this cannot be the percentage we require — it obviously should be more than 50 per cent. So the proportion of bags meeting the claim is $100 - 6.68 = 93.32$ per cent.

Example 3

Sometimes it is necessary to use the normal distribution in reverse, i.e. given a proportion, we enter the tables to find the corresponding Z score, and so find a corresponding value in a given normal distribution. Suppose that in order to satisfy legal requirements, a pork pie manufacturer may only

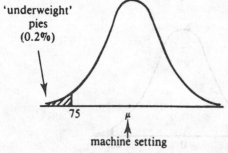

'underweight' pies (0.2%)

75   μ

machine setting

Figure 3.5

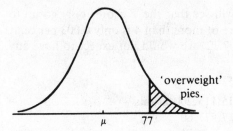

Figure 3.6

produce 0.2 per cent of pies below a weight of 75 grams. The pie-producing machine operates with a standard deviation of 0.5 grams. If the weights of pies are normally distributed, to what (mean) weight should the machine be set? Fig. 3.5 illustrates the situation, and μ is the machine setting required.

Consulting the table we see that the nearest Z score to a proportion 0.2 per cent is 2.88. Notice that the weight we are considering (75 grams) is *less than* the mean, so its Z score must be negative. In other words,

$$\frac{75 - \mu}{0.5} = -2.88$$

so the machine setting (μ) is 76.44 grams (i.e. $2.88 \times 0.5 + 75$).

Let us take this example a stage further. The pie manufacturer's weekly output is 200,000 pies, and the pie contents cost 5p per 100 grams. Pies with weights in excess of 77 grams require additional packing at a cost of 0.5p per pie. Find the firm's weekly cost.

First, we are going to have to find the proportion of 'overweight' pies. This is illustrated in Fig. 3.6. The Z score is

$$\frac{77 - 76.44}{0.5} = 1.12$$

so 13.14 per cent of output will need extra packaging.

We can now calculate the weekly cost like this:

weekly pie content requirement is $200,000 \times 76.44 = 15,288,000$ grams
weekly cost of contents is $15,288,000 \times 0.05\text{p} = £7,644$
each week 13.14% of $200,000 = 26,280$ pies need extra packing
weekly cost of extra packing is $26,280 \times 0.5\text{p} = £131.40$

Finally, let us suppose that a new pie-producing machine is available, operating with a standard deviation of 0.2 grams but costing an additional £225 per week to operate. Would you recommend that the pie manufacturer purchases the new machine? To answer this question, we must first determine the machine setting using the new machine, and determine the percentage of 'overweight' pies. To obtain the machine setting, we know that:

$$\frac{75 - \mu}{0.2} = -2.88$$

so the machine setting (μ) must be 75.576 grams. (Notice that the lower standard deviation enables the machine setting to be lowered.) To obtain the percentage of 'overweight' pies, we know that:

$$Z = \frac{77 - 75.576}{0.2} = 7.12$$

The normal distribution 39

Now if you consult the normal distribution tables, you will see that the Z scores only go up to 4.0, and the proportion of normal distribution with a Z score of more than 4 is only 0.003 per cent. Clearly, then, it follows that as 77 grams has a Z score or 7.12, we would not expect to have any 'overweight' pies if the new machine is used.

We can now calculate the cost of using the new machine:

weekly pie content requirement is $200,000 \times 75.576 = 15,115,200$ grams

weekly cost of contents is $15,115,200 \times 0.05\text{p} = £7,577.60$

and as there are no 'overweight' pies, this is the total weekly cost of using the new machine. So the new machine involves a weekly cost saving of $£7,775.40 - £7,557.60 = £217.80$. However, as the new machine costs an extra £225 per week to operate, the manufacturer would be well advised to retain the use of the old machine.

3.2.1 The normal approximation to the binomial distribution

So far we have used the normal distribution to find the proportion in a population that was greater than or less than a given value. Now in fact, the normal distribution is another type of probability distribution. Earlier, we asked what *proportion* of electric lamps failed before (say) 1,000 hours but in fact we were also finding the probability that a lamp failed before 1,000 hours. Can you see why? Surely, the proportion of lamps failing before 1,000 hours can be taken as a measure of the probability that a particular lamp fails before this period. So when we are considering a normal distribution, *we can consider the words proportion and probability to be interchangeable*. The binomial distribution and the normal distribution, then, are two types of probability distributions. However, there is a subtle difference between them – the normal distribution is continuous in so far as the measurements considered can have any value (they are a continuous variable). The binomial distribution is discrete (we are concerned with integers only).

The greatest problem in using the binomial distribution is that it involves awkward arithmetic. Now if the sample size is very large, we can use the normal distribution as an approximation to the binomial distribution, and this will save a considerable amount of arithmetic. Suppose we draw a sample of 1,000 packets of soap powder from a batch in which 20 per cent of items are underweight, and we require to know the probability of obtaining more than 220 underweight packets in our sample. The appropriate binomial distribution is $(0.8 + 0.2)^{1,000}$, and the probability we require is $P_{(221)} + P_{(222)} + P_{(223)} + \cdots + P_{(1,000)}$. Obviously this is going to be quite some task! Now the appropriate normal distribution to use is the one with the same mean and standard deviation as the binomial distribution. In this case, we have:

$$\bar{x} = np = 1,000 \times 0.2 = 200$$

$$\sigma = \sqrt{npq} = \sqrt{1,000 \times 0.2 \times 0.8} = 12.65$$

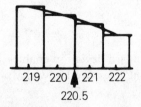

Figure 3.7

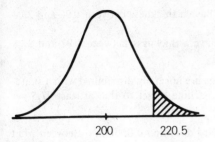

Figure 3.8

In Fig. 3.7 we have drawn part of the histogram of the probability distribution $(0.8 + 0.2)^{1,000}$ and it is *not* drawn to scale. If we want to find the probability of obtaining more than 220 packets, we must add the areas of the rectangles enclosing 221, 222, 223,..., 1,000.

The line joining the mid-points at the top of each rectangle represents the normal distribution that we are using as an approximation. If we wish to use this normal distribution as an approximation, then the diagram clearly shows that we require the proportion to the right of 220.5. This 0.5 adjustment is always used when we use the normal distribution as an approximation to a discrete distribution. So the situation looks like that shown in Fig. 3.8

The Z score for 220.5 is

$$\frac{220.5-200}{12.65} = 1.62$$

and consulting the normal distribution tables we see that the probability we require is $100-94.74 = 5.26$ per cent.

Before ending this chapter, two points deserve attention. First, the normal distribution is of paramount importance in statistics, and you would be well advised to master it before moving on to the following chapters. Secondly, you may be rather worried by the stringent conditions imposed before the normal distribution may be used. Well, there are many cases in which a population shows a very good approximation to the normal distribution, and any errors resulting from its application should not be too serious. More importantly, although populations may not approximate to the normal distribution, there are sound theoretical reasons for assuming that sampling statistics will do so. It is to this important truth that we shall shortly turn our attention.

Review question

3.1 The average number of newspapers purchased by a population of urban households in 1972 was 400, and the standard deviation was 100. Assuming that the purchases were normally distributed, find the proportion of households that bought:

    (a) between 250 and 500 papers,
    (b) less than 250 papers,
    (c) between 500 and 600 papers,
    (d) more than 600 papers.

3.2 Assuming that the hub thickness of a certain type of gear is normally distributed with a mean 2.00 inches, and a standard deviation of 0.04 inches

(a) how many gears in a production run of 5,000 gears will have a thickness between 1.96 and 2.04 inches?

(b) how confident can you be that an individual gear will have a thickness between 1.92 and 2.08 inches?

3.3 An automatic packaging machine produces packages whose weights are normally distributed with a standard deviation of 0.07 ounces. To what mean weight should the machine be set so that at least 97.5 per cent of packages are over 1 lb in weight?

3.4 A machine cuts copper pipes to a mean length of 2 metres, standard deviation 0.05 metre. Between what limits would you expect the lengths of 95 per cent of pipes to lie?

3.5 A firm is considering the purchase of a machine to turn ball-bearings. A machine is borrowed for testing purposes, and it is found that 6.68 per cent of ball-bearings have mean diameters greater than 5.03 mm.

(a) Assuming that the machine is set at 5 mm, find the standard deviation to which the machine operates.

(b) The firm wishes to produce ball-bearings with diameters within 0.05 mm of a nominal 5 mm, and any outside this range would be rejected. Assuming the machine produces 1 million units per month, and and that each unit rejected costs the firm 1p, find the monthly cost of rejects.

(c) A more accurate machine is available, which costs an additional £100 per month to buy and operate. If this operates with a standard deviation of 0.016 mm, which machine should be purchased?

3.6 This final question will test your knowledge on all three probability distributions.

Among the products of a certain lumber company are beadings of mean length 80 cm and standard deviation 2 cm. The production process is estimated to produce 10 per cent 'misshaped' beadings. It is specified also that the average number of knots per 100 cm of beading is 1.5.

(a) If one beading is chosen, what is the probability that its length is between 83 cm and 85 cm?

(b) If a sample of five beadings is selected, what is the probability that there are three misshapes?

(c) What is the probability that one beading of mean length will contain two or less knots?

4 Decision criteria: expectation

4.1 Introduction

In this chapter we will apply probability theory to the decision-taking process. We are all experts in taking decisions. Does this surprise you? Well throughout your life you have had many decisions to take, and the fact that you are alive today, reading this book indicates that you have taken many decisions correctly. Take a simple problem, such as the need to get to the other side of a busy road. You have to decide when it is safe to cross. The fact that you have so far taken the correct decision proves that you have carefully weighed up the evidence, and interpreted it correctly. Sometimes the decision will be easy to take: the road might be completely empty and you can cross with safety; and at other times the road will be so busy that to attempt to cross would be suicidal. In other words, we have all the necessary information available to make the correct decision every time. A cautious (and according to road safety officers a sensible) person would obey the following decision rules: Only cross the road when no traffic is in sight. If we obey this rule, then we will always cross the road safely (unless we fall down an uncovered manhole!). Unfortunately, however, it is not always possible to obey this rule, as there are many roads so busy that we could never cross. With normal city centre roads we will seldom find a situation where they are completely empty of traffic, and we must use our judgement to decide when it is safe to cross.

Now, of course, the decision-taking problems faced by business people are much more complex than the decision whether to cross the road. Not only must they decide between many alternatives, but also they often cannot be sure of the consequences of any decision made. In this chapter, we shall attempt to discover criteria that may aid the decision-taking process.

4.2 Decision criteria

Guy Rope owns a camp site in the Dordogne, and wishes to develop it in order to increase profits. He realises that three options are open to him: he could build a swimming pool and charge the campers for its use; he could build a tennis court and charge for its use; or he could build a restaurant to supply the campers with simple hot meals. He has sufficient funds to undertake just one of the options – what should he do? Obviously, the decision taken by Guy will depend on the profitability of the options and he realises that the profitability will depend upon the weather. If the summer is too hot, then tennis may prove to be too exhausting; also evening barbecues in the open air will be more popular than restaurant meals. Given a hot summer, then, the swimming pool would be the most profitable option. If the summer is poor, then Guy reckons that the campers would welcome somewhere where they could buy hot meals, and so the restaurant would be the most profitable.

People seem to prefer playing tennis when the weather is neither too hot nor too cold, so the tennis court would prove to be the most profitable given an 'average' summer. Armed with these beliefs, Guy estimates the annual profitability of each course of action as in Table 4.1 (all figures in thousand francs).

Table 4.1

| | | State of nature Summer is | | |
		Cool	Average	Hot
	Swimming pool	50	100	150
Strategies	Tennis court	30	180	90
	Restaurant	170	100	40

Notice that the options facing Guy are called *strategies*. A list of strategies is a list of courses of action facing the decision taker who has direct control over which course is chosen. However, Guy has no control whatsoever over the weather (if he had, then there would be no problem!). The different types of summer that can occur are called *states of nature*, a list of events outside the control of the decision taker. The table above is usually called a *payoff matrix*. For each strategy it shows the payoff (in this case, the profit) that would result from each state of nature.

What should we advise Guy to do? Unfortunately, mathematics alone cannot help us – we need to know something about Guy's character. What sort of person is he? Is he a gambler, a risk taker who is supremely confident that lady luck is always on his side? If Guy is this type of person, then he would reason as follows: 'Lucky me – I am always right. If I build a swimming pool, then the summer will be hot and I will earn 150,000FF. If I build a tennis court, then the summer will be average and I will earn 180,000FF. If I open a restaurant, the summer will be cool and I will earn 170,000FF. The most logical thing for me to do is to go for the greatest profit, so I will build a tennis court.'

Table 4.2 summarises Guy's reasoning. For each strategy, he notes the maximum payoff that can result. He then chooses the strategy with the greatest maximum payoff. If Guy acts in this way, then he is applying the *maximax criterion*: he is maximising his possible payoff by gambling that the summer will be average.

Table 4.2

	Cool	Average	Hot	Maximum payoff	
Swimming pool	50	100	150	150	
Tennis court	30	180	90	180	←
Restaurant	170	100	40	170	

Decision takers who practice the maximax criterion must be very rare specimens. If maximax was widely practised, then although we would have a few more millionaires, we would certainly have many more bankruptcies! Lady luck smiles on very few of us indeed! We shall now move to the other end of the spectrum and suppose that Guy is a born pessimist, who always assumes that the states of nature work against him. He would argue something like this: 'Suppose I build a swimming pool. Will the summer be hot or even average? No chance! You can bet your bottom dollar that the summer will be cool! Likewise, one sure fire way to ensure a cool summer is for me to build a tennis court. Of course, if I open a restaurant then the summer will be the hottest for years and only a few campers will want feeding. The big money always avoids me. I might as well build a swimming pool as this at least guarantees me 50,000FF profit. If I build a tennis court I could only be sure of 30,000FF and opening a restaurant guarantees me no more than 40,000FF. A swimming pool it is, then.' Table 4.3 summarises Guy's reasoning.

Table 4.3

	Cool	Average	Hot	Minimum payoff
Swimming pool	50	100	150	50 ←
Tennis court	30	180	90	30
Restaurant	170	100	40	40

For each strategy, he notes the minimum payoff that can result. He then selects the strategy that maximises his minimum payoff. Guy is applying the *maximin criterion*; he is maximising his minimum possible payoff.

So we see that the two criteria are fundamentally different, reflecting quite different states of mind. If Guy adopts the maximax criterion, then he is pushing the ceiling on his profits to the highest level (180,000FF) by building a tennis court. No other strategy could earn this much. However, he is taking a risk: his profit would only be 30,000FF if the summer was cool. On the other hand, if Guy adopts the maximin criterion, then he is raising the floor on his profits to the highest level (50,000FF) by building a swimming pool. No other strategy could guarantee as much as this. However, by building a swimming pool he forgoes the possibility of really large profits (180,000FF from a tennis court or 170,000FF from a restaurant).

There is a third way of analysing the problem facing Guy – a way that neither assumes that Guy is ultra optimistic nor assumes that he is ultra pessimistic. The reasoning is something like this: suppose the summer turns out to be cool – if Guy had opened a restaurant then he would have made the right decision. He would have no 'regret' at all, and earn the maximum possible profit under the circumstances (170,000FF). But suppose he had built a swimming pool – he has made the wrong decision and would certainly regret it. His profit would be 50,000FF, 120,000FF less than it would have been had he made the correct decision. It would seem reasonable, then, to use the 120,000FF as a measure of the extent of his regret. In a similar fashion, if the summer was cool and Guy had built a tennis court, then he would regret the $170,000 - 30,000 = 140,000$FF he had forgone. If we use similar reasoning and assume an average summer, then assume a hot summer, we can calculate Guy's regret under all possible circumstances. Our results are summarised in a *regret matrix* (see Table 4.4).

Table 4.4

	Cool	Average	Hot	Maximum regret
Swimming pool	120	80	0	120
Tennis court	140	0	60	140
Restaurant	0	80	110	110 ←

For each strategy, the maximum regret has been identified (for example, had he built a tennis court, then Guy's maximum regret would have been the 140,000FF in profit forgone during a cool summer). Examining the matrix, we conclude that Guy should open a restaurant, as this is the strategy that minimises the maximum regret he could experience. If he uses this reasoning as a basis for decision taking, then Guy is applying the *minimax criterion*.

In this section, we have examined three quite distinct criteria for decision taking, and we shall now state a few simple rules to summarise them. The rules will assume that in the payoff matrix the rows refer to the strategies and the columns refer to states of nature.

Rule 1

For each row in the payoff matrix, find the maximum value. If we select the row with the greatest maximum value, then we are applying the maximax criterion.

	Cool	Average	Hot	Maximum	
Swimming pool	50	100	150	150	
Tennis court	30	180	90	180	←
Restaurant	170	100	40	170	

Rule 2

For each row in the payoff matrix, find the minimum value. If we select the row with the greatest minimum value, then we are applying the maximin criterion.

	Cool	Average	Hot	Minimum	
Swimming pool	50	100	150	50	←
Tennis court	30	180	90	30	
Restaurant	170	100	40	40	

Rule 3

Find the maximum value for each column in the payoff matrix, and subtract all the payoffs in each column from their corresponding maximum value. This gives the regret matrix.

Rule 4

For each row in the regret matrix, find the maximum value. If we select the row with the least maximum value, then we are applying the minimax criterion

	Cool	Average	Hot
Swimming pool	50	100	150
Tennis court	30	180	90
Restaurant	170	100	40
Maximum	170	180	150

	Cool	Average	Hot	Maximum regret	
Swimming pool	120	80	0	120	
Tennis court	140	0	60	140	
Restaurant	0	80	110	110	←

4.3 Which criteria?

We have applied three different criteria to the problem facing Guy. The maximax criterion suggests he should build a tennis court. The maximin criterion suggests he should build a swimming pool.

The minimax criterion suggests he should open a restaurant. Poor Guy must be utterly confused! Which criterion is appropriate to this problem? As we suggested earlier, this is an impossible question to answer as the strategy will depend upon Guy's character. However, it is possible to comment generally on the criteria.

(a) For the problem facing Guy, each criterion suggested a different strategy, but it frequently happens that different criteria would suggest the same strategy. For example, if the payoff matrix was:

	Cool	Average	Hot
Swimming pool	30	100	170
Tennis court	50	180	90
Restaurant	170	100	40

then whatever criterion is applied the strategy selected would be the same – build a tennis court (you should verify for yourself that this is true). This strategy is said to be *dominant*, and no problem of strategy selection exists.

(b) The minimax criterion minimises the decision taker's maximum regret, and so it attempts to minimise the consequences of taking the wrong decision. This is an intrinsically satisfying criterion – it seems to be a highly logical method of decision taking. Moreover, this criterion fits in well with economists' ideas of opportunity cost (if you do not know what opportunity cost is, talk to an economist).

(c) If we examine the payoff matrix facing Guy, then we notice that whatever strategy he selects and whatever the state of nature, he always makes a profit. Because of this, Guy may be tempted to apply the maximax criterion – especially if he is what economists call a profit maximiser. He would be following the well quoted 'law' that you must speculate to accumulate. But suppose it was possible for losses to occur: this could well be the case if he opened a restaurant and the summer was hot, or he built a swimming pool and the summer was cool. Surely, Guy would then be strongly tempted to apply the maximin criterion, especially if the losses could threaten his survival as a camp site owner.

(d) Guy has a problem selecting the appropriate strategy because he has no control over the state of nature, but if he has *more information* about the states of nature, then his decision can be more soundly based. What additional information would help? Well, the states of nature will not be equally likely to occur, so if Guy can assign probabilities to them this should increase his insight into the problem.

4.4 Expected monetary value

Suppose that Guy examines the Dordogne weather records over the last 100 years. On 20 occasions the summer was cool, on 70 occasions it was average and on 10 occasions it was hot. This information enables Guy to deduce the following empirical probabilities:

State of nature	Probability
Cool	0.2
Average	0.7
Hot	0.1

Suppose Guy thinks ahead over the next ten years. Using the probability distribution he would predict two cool, seven average and one hot summer. So if he was to build a swimming pool, he would estimate his earnings (in thousands francs) at:

$2 \times 50 + 7 \times 100 + 1 \times 150 = 950$

which is an average of $950 \div 10 = 95$ per year. This average earning is called the *expected monetary value* (EMV) of building a swimming pool, and we could calculate it more directly like this:

$0.2 \times 50 + 0.7 \times 100 + 0.1 \times 150 = 95.0$

In a similar fashion we could deduce that:

EMV (build tennis court) $= 0.2 \times 30 + 0.7 \times 180 + 0.1 \times 90 = 141$
EMV (open restaurant) $= 0.2 \times 170 + 0.7 \times 100 + 0.1 \times 40 = 108$

We now have another criterion for selecting the appropriate strategy: *select the strategy with the greatest EMV* as this maximises the long-run average gains. Guy would be well advised to build a tennis court and earn an average of 141,000FF per year.

Of the four criteria we have examined, EMV has the soundest logical base. Maximising average long-run profits certainly seems a sensible thing to do. However, the criterion does have its drawbacks. We have stated that Guy should build a tennis court because this has the greatest EMV (141,000FF) – but on any particular year he cannot earn this amount. He will earn either 30,000FF or 180,000FF or 90,000FF. Suppose one of these options was a loss, then we would have to enquire whether Guy is capable of shouldering the loss. Sooner or later, losses would be bound to occur, and if they could result in bankruptcy, then Guy is clearly using the wrong criterion. The maximin criterion would be more appropriate.

The second problem with using the EMV criterion arises when the term 'long-run payoff' makes no sense. In particular, let us suppose that Guy intends to run the camp site for just one summer, obtain the maximum profit he can, then leave the industry (he would be what economists call a 'snatcher'). The EMV of building a tennis court cannot be his long-run average profit as there will be no long run! His average profit will be the profit for the year – either 30,000FF or 180,000FF or 90,000FF. But if he stays in the industry, then as the years elapse, the closer will his average annual profit move towards 141,000FF. If Guy is a snatcher, must we conclude that the EMV criterion is inappropriate? Not necessarily. If Guy applies the EMV criterion to all the decision problems facing him, however diverse they may be, then he would be maximising his returns over the entire range of problems.

4.5 The value of perfect information

For the rest of this problem, we will assume that Guy is a snatcher, basing his decisions on the EMV criterion. He decides to build a tennis court as this has the greatest EMV (141,000FF). Now suppose additional information is available to Guy – in particular, let us suppose that a peasant is prepared to predict the weather and that his predictions are always right. In the jargon, we would say that Guy has *perfect information* available to him. This will certainly help the decision-taking process. If the peasant predicts a cool summer (and there is a 20 per cent chance that he will) then Guy will open a restaurant. If he predicts an average summer (and there is a 70 per cent chance that he will) then Guy will build a tennis court. If he predicts a hot summer (and there is a 10 per cent chance that he will) Guy will build a swimming pool. Given perfect information, then, Guy will earn either 170,000FF or 180,000FF or 150,000FF. His expected earnings, then, are:

$0.2 \times 170 + 0.7 \times 180 + 0.1 \times 150 = 175,000FF$

Without this information, Guy's expected earnings were 141,000FF. So the information has a value to Guy of 175,000 − 141,000 = 34,000FF. This 34,000FF is the *expected value of perfect information* (EVPI). Now it is highly likely that the peasant will charge for the information, but it would certainly be worth buying if its cost was less than EVPI. In other words, if the peasant charges less than 34,000FF for his predictions, then they are worth buying, but if the peasant charges more than 34,000FF, then Guy would be advised not to buy and simply apply the EMV criterion.

4.6 Multi-stage decision analysis

We shall now examine a more complex problem, in which the decision-taking process is divided into stages. Flint McRae owns the oil prospecting and development rights for a plot of land in California, and a property development company has offered him $50,000 for the plot. Flint must now decide whether to sell the land or to exploit it as an oilfield. Suppose that Flint decides to exploit the land – he sinks the well but finds no oil. In the jargon of the oil industry, the well is said to be 'dry', and Flint would lose $20,000. However, the well might yield oil in commercial quantities – a so-called 'wet' well – and this would earn Flint $100,000 in total income. Exceptionally, Flint could obtain a real bonanza from his oilwell, a so-called 'soaking', well, and this would earn $500,000 in total net income. From past experience, Flint knows that there is a 70 per cent chance that the well is 'dry', a 20 per cent chance that it is 'wet' and a 10 per cent chance that it is 'soaking'. If he wishes, Flint can engage a firm of geologists to undertake a seismic survey of the land, and this will cost him $30,000. The geologists supply the following information as to the reliability of such surveys (all figures are prior probabilities).

		Survey report is		
		Bad	Good	
	Dry	0.44	0.26	0.7
True state of the well	Wet	0.05	0.15	0.2
	Soaking	0.01	0.09	0.1
		0.50	0.50	1.0

If Flint always uses the EMV criterion, what should he do?
In this problem, there are two sets of strategies facing Flint:

(a) Should he engage the geologists' services?
(b) Should he drill or should he sell?

Likewise, there are two sets of states of nature facing Flint:

(a) The result of the survey (is the report good or bad?).
(b) The true state of the well (is it dry, wet or soaking?).

Armed with this information, we can construct a tree diagram of the problem facing Flint (Fig. 4.1). The boxes represent decision points and the circles represent states of nature. So we see that there are four distinct decision points facing Flint.

Now before Flint can decide on decision point 1, he must decide on decision points 2, 3 and 4. Hence we will analyse the tree diagram from right to left – a method called the *rollback principle*.

First we shall analyse decision point 2 (Fig. 4.2).
Assuming that Flint reaches decision point 2:

EMV (sell) = $50,000
EMV (drill) = 0.7 × − $20,000 + 0.2 × $100,000 + 0.1 × $500,000
= $56,000

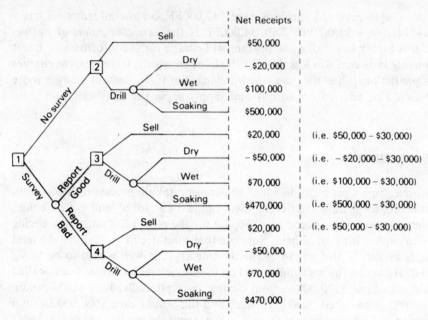

Figure 4.1

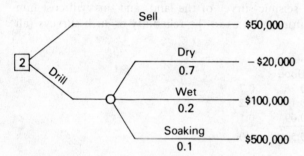

Figure 4.2

So we see that if decision point 2 reached, then Flint will drill. If we now turn to decision point 3 (see Fig. 4.3), we will need conditional probabilities, i.e. the probabilities that the well is dry, wet or soaking given a good report from the geologist.

P (dry well|good report) $\quad = 0.26 \div 0.5 = 0.52$
P (wet well|good report) $\quad = 0.15 \div 0.5 = 0.30$
P (soaking well|good report) $= 0.09 \div 0.5 = 0.18$

Assuming that Flint reaches decision point 3:

EMV (sell) $= \$20,000$
EMV (drill) $= 0.52 \times -\$50,000 + 0.3 \times \$70,000 + 0.18 \times \$470,000$
$\qquad\qquad = \$79,000$

So if decision point 3 is reached, Flint should drill. Turing now to decision point 4 (Fig. 4.4):

P (dry well|bad report) $\quad = 0.44 \div 0.5 = 0.88$
P (wet well|bad report) $\quad = 0.05 \div 0.5 = 0.10$
P (soaking well|bad report) $= 0.01 \div 0.5 = 0.02$

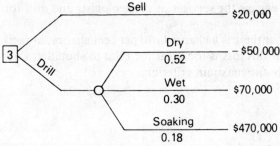

Figure 4.3

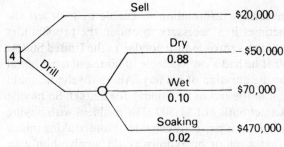

Figure 4.4

Assuming that Flint reaches decision point 4:

EMV (sell) = $20,000
EMV (drill) = 0.88 × − $50,000 + 0.1 × $70,000 + 0.02 × $470,000
= − $27,600

Clearly, then, if decision point 4 is reached, then Flint should sell.

Now that we have evaluated decision points 2, 3 and 4, we can evaluate decision point 1. Before we do this, however, it might be useful to summarise what we have concluded. If Flint does not employ the geologists' services, then his best course of action is to drill (EMV = $56,000). If he does employ the geologists' services, then his action will depend upon the geologists' report. If the report is good, then Flint should drill (EMV = $79,600) but if it is bad he should sell (EMV = $20,000).

Evaluating decision point 1 (Fig. 4.5) we have:

EMV (no survey) = $56,000
EMV (survey)   = 79,600 × 0.5 + 20,000 × 0.5 = $49,800

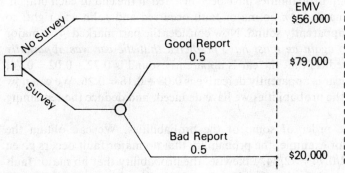

Figure 4.5

Finally, then, we conclude that Flint should not engage the services of the geologists and drill for oil.

Before leaving this problem, you should notice that there is a chance (a 70 per cent chance, in fact) that Flint faces a loss of $20,000 if he drills. Now if this loss is too great for Flint to shoulder, then he should sell to the property developer, i.e. apply the maximin criterion.

4.7 Revising probabilities

In the previous example, we were given all the necessary information to enable us to insert the probabilities directly on to the decision tree. Sometimes it is necessary to obtain the probabilities ourselves. Consider the case of John Smith, who is taking a three-month holiday in the United States. He realises that his holiday would be more enjoyable if he had a car available to him, and a car rental firm quotes him a fee of $2,500 to rent him a car for the duration of his stay. Alternatively, he could buy a used car of the same type for $3,500, and sell it at the end of his holiday for $2,000. So buying rather than renting would involve John Smith with a net outlay of $1,500. The problem with buying a used car is that the car might break down, and while John is perfectly capable of undertaking minor repairs himself at a negligible outlay, he reckons that a major breakdown could involve him with repair charges of $1,500. A friend advises John that the probability that a car of this age and type will break down over the next three months is 20 per cent. There is a further alternative facing John Smith: he could persuade his friend to rent the used car for a day for $50. John and his friend could then examine the vehicle, and assess its likelihood of breaking down over the next three months. There is a 90 per cent chance that any prediction they make will be right. What should John Smith do?

First we shall draw the decision tree of the problem, inserting the associated costs (Fig. 4.6). We know that the probability of no major fault occurring is 0.8, so we can evaluate decision point 2:

cost if car rented = $2,500
expected cost if car bought = $1,500 \times 0.8 + 3,000 \times 0.2 = \$1,800$

So if decision point 2 is reached, John Smith would buy the car as this action has the lower expected cost.

Evaluation of decision points 3 and 4 is not so straightforward. We need to know the probabilities that the car is apparently sound and apparently defective, and Fig. 4.7 illustrates the problem.

The first branch of the tree diagram gives the true experience with the car: either a major fault occurs or it does not. Each of the branches now divide according to whether or not John and his friend made a correct prediction. The joint probabilities have been inserted at the end of each branch. Consider the part of the tree diagram marked ①: no major fault occurred and John was right, so he must have predicted that the car was apparently sound. Now consider the part marked ④: a major fault occurred and John was wrong, *so again he must have predicted that the car was apparently sound*. So, the probability that John predicts that the car is apparently sound is $0.72 + 0.02 = 0.74$, and the probability that he says that the car is apparently defective is $0.08 + 0.18 = 0.26$. We can now draw the tree diagram in reverse, insert the probabilities we have deduced, and deduce the remaining probabilities (Fig. 4.8).

Notice that is necessary to alter the order of some of the probabilities. We can obtain the conditional probabilities by division – for example, the probability that no major fault occurs given that John thinks the car is sound is $0.72/0.74 = 72/74$. Likewise, the probability that no major fault occurs given that John thinks the car is defective is $0.08/0.26 = 8/26$. We can now evaluate the

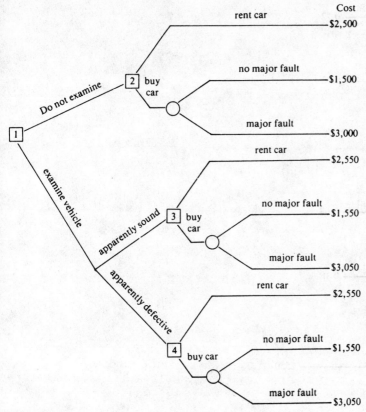

Figure 4.6

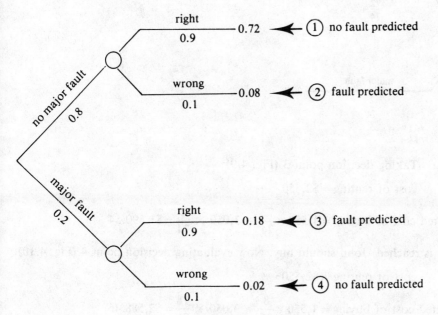

Figure 4.7

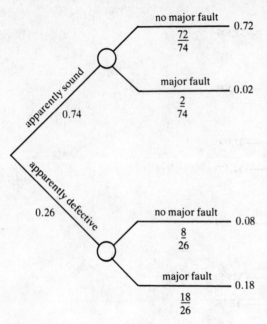

Figure 4.8

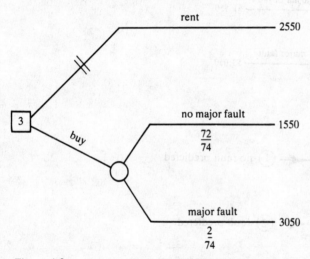

Figure 4.9

remaining decision points. Taking decision point 3 (Fig. 4.9):

$$\text{cost of renting} = \$2,550$$

$$\text{expected cost of buying} = 1,550 \times \frac{72}{74} + 3,050 \times \frac{2}{74} = \$1,590.54$$

So if decision point 3 is reached, John should buy. Now evaluating decision point 4 (Fig. 4.10):

$$\text{cost of renting} = \$2,550$$

$$\text{expected cost of buying} = 1,550 \times \frac{8}{26} + 3,050 \times \frac{18}{26} = \$2,588.46$$

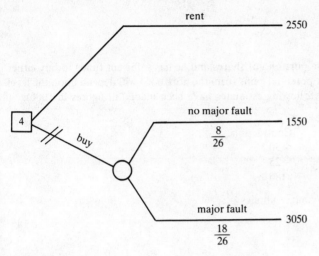

Figure 4.10

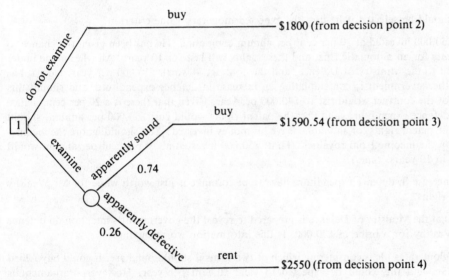

Figure 4.11

So if decision point 4 is reached, John should rent. Finally, we can evaluate decision point 1 (Fig. 4.11):

expected cost if vehicle not examined = $1,800
expected cost if vehicle examined = $1,590.54 \times 0.74 + 2,550 \times 0.26 = $1,840$

We would advise John to go ahead and buy the vehicle outright, though the decision is very marginal. Should the cost of borrowing the car for a day be less than $50, and/or should John and his friend be more accurate in their prediction, then the alternative course of action might well be chosen.

Review questions

4.1 An investment trust manager wishes to buy a portfolio of shares and he has sufficient funds to buy either Portfolio A, Portfolio B or Portfolio C. The potential gains from the portfolios will depend upon the level of economic activity in the future, and the following estimates have been made (all figures in £000):

	State of nature		
Portfolio	Expansion	Stability	Contraction
A	100	50	− 50
B	50	100	− 25
C	− 50	0	180

Which portfolio should be selected if the manager applies:

    (a) The maximax criterion?
    (b) The maximin criterion?
    (c) The minimax criterion?

4.2 Suppose that in question 4.1 the investment manager makes the following probability estimates:

Expansion	Stability	Contraction
0.1	0.4	0.5

Which portfolio should be selected if he uses the expected monetary value criterion?

4.3 Mr Wealthy has £50,000 invested at 10 per cent per annum compound. He has been given the chance to buy the patent rights for an automatic rifle, and these rights will last for 10 years. All rifles made under this patent are sold to the Ministry of Defence, and the contract is worth £10,000 per year. Now it has been hinted that the government is contemplating an increase in defence expenditure, and should this happen the value of the contract would rise to £40,000 per year. Given that there is a 20 per cent chance of an increase in defence expenditure, and that the patent rights would cost £50,000 per annum, should Mr Wealthy buy the patent rights, or should he leave his money invested? You should ignore the income-earning capacity on the income from royalties. (Hint: £50,000 invested at 10 per cent per annum would grow to £129,687 in 10 years' time.)

4.4 What would the increase in defence expenditure have to be to make it just worth while for Mr Wealthy to buy the patent rights?

4.5 Suppose an official at the Ministry of Defence is prepared to reveal the government's intention on defence spending to Mr Wealthy for a bribe of £30,000. Is this information worth buying?

4.6 International Conglomerates plc is considering which of two firms it should purchase. It could buy Allied Dog Foods for £3m and this could be expected to yield £0.75m per year. However, Parliament is considering legislation to restrict the number of dogs. A decision is not expected for 2 years, but if the number of dogs is restricted, then the annual receipts would fall to £0.1m. Alternatively, Guided Systems could be bought for £1m, and this would yield a cash flow of £0.5m per year. Within 2 years, the government must decide whether to replace the present early warning system, and Guided Systems must decide whether or not to expand at a cost of £2m and tender for the replacement system. If the tender is successful, Guided Systems' cash flow would increase to £0.8m per year, but if the tender is unsuccessful (or if no tender is made) then it would fall to £0.1m per year. Any decision taken now must last for the next 10 years.

    Draw a tree diagram of the alternative courses of action facing International Conglomerates. Insert the payoffs and decide on the best course of action under the maximax and maximin criteria.

4.7 For the last question, find the best course of action according to the EMV criterion given the following probabilities:

P (government restricts number of dogs) = 0.4
P (government replaces early warning system) = 0.7
P (Guided Systems' tender is successful) = 0.9

4.8 If an official at the Home Office is prepared to release information on the government's intention towards restricting the number of dogs, how much is this information worth to International Conglomerates?

4.9 A manufacturer has spent £20,000 on developing a product, and must now decide whether to manufacture on a small scale or a large scale. If the demand for the product is high, then the expected profit during the product's life would be £700,000 for a high manufacturing level and £150,000 for a low manufacturing level. If demand is low, then the expected profit is £100,000 for a high manufacturing level and £200,000 for a low manufacturing level. The initial indication is a 40 per cent chance of a high demand, but a market research survey could predict the demand with 85 per cent accuracy. How much can the manufacturer afford to spend on market research?

Part Two   Elementary forecasting

5   The time series

5.1   Introduction

In this chapter, we will take the first steps in analysing data that varies over time. You will be shown how to calculate the trend of such data and, where appropriate, measure the strength of any seasonal forces at work. We can then eliminate seasonal forces so as to obtain a meaningful picture of performance. Other factors affecting data over time, such as cyclical and residual influences, will be discussed. Finally, you will be introduced to an elementary method of statistical forecasting.

5.2   The components of a time series

If Plumpton Rovers were drawn against Arsenal in the third round of the FA Cup, there is no doubt who would be expected to win. Past records show that non-league clubs seldom win against First Division teams. If, however, the non-league club had won such a tie on eighteen out of the last nineteen meetings, then in spite of their relative strength we would expect Arsenal to lose. We tend to base our forecast of the match result on what has happened in the past. This is as true of the accountant as it is of the football fan. In trying to budget for the cost of raw materials or for labour cost in the immediate future, he/she will look carefully in the first instance at what has happened in the recent past, expecting that the past behaviour of costs will be carried forward into the future.

The statistical series which tells us how data has been behaving in the past is the *time series*. It gives us the value of the variable we are considering at various points in time: each year for the last 15 years; each quarter for the last 5 years. Yet, when you look at a typical time series, the data fluctuates so much that it seems unlikely that it can help us a great deal. Let us take a series, then, and begin by assessing the factors which cause this fluctuation in the value of the data.

Imports of raw material into Ruritania
(19-0 = 100)

Year	Quarter			
	1	2	3	4
19-2	114	142	155	136
19-3	116	150	153	140
19-4	128	158	169	159
19-5	137	180	192	172
19-6	145	194		

The figures are given in index number form, and they are plotted as a graph in Fig. 5.1.

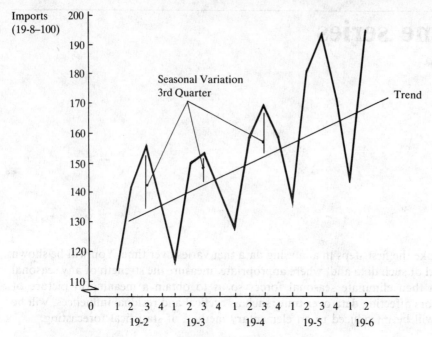

Imports (19-8–100)

Figure 5.1   Imports of raw materials into Ruritania

Look carefully at Fig. 5.1. Although the figures are fluctuating quarter by quarter, two things are immediately apparent. There is a general upward *trend* in the figures as a whole. It is not an exceptional rise, but it is quite marked. Apart from the third quarter of 19-3, the figures in any quarter are higher than those in the same quarter of the previous year.

Secondly, although the figures fluctuate, there is a pattern. The index of imported raw materials is always highest in the third quarter of the year, and always lowest in the first quarter. There is, then, a very marked *seasonal fluctuation* in the figures.

Although it is not apparent in the graph, there is likely to be a third influence on the data. We would not expect import figures to remain unaffected by day-to-day happenings such as disputes in the docks, exchange rate fluctuations or fuel shortages. It is probably something of this nature that has reduced the figure for the third quarter of 19-3. Thus, our data is likely to be affected by what we can call *residual* or *random influences* which cannot be foreseen or pinpointed without a great deal of outside knowledge – but which may be of great importance.

The series we are looking at is, of course, a short one. We could link the trend perhaps to a general expansion of the economy during the upward phase of the trade cycle. If we were to extend the series and consider a far longer period, the upward expansion might reach a peak and the trend would begin to turn downwards as the economy moves towards depression. Thus, the series we examine might be affected to a great extent by *cyclical influences* resulting in a trend such as that in Fig. 5.2.

Thus, data such as this which fluctuates quite markedly over time may be responding to any, or all, of four sets of forces:

(a) The *trend* or general way in which the figures are moving. It is important to distinguish between a trend resulting from cyclical influences on the economy and one resulting from, for example, a change in tastes or consumer buying habits. It is likely that there would have been a downward trend in the sales of black and white television sets, even though the economy as a whole was expanding and the sales of all television sets (including colour) were rising.

(b) *Seasonal variations*. It is common knowledge that the value of many variables depends in

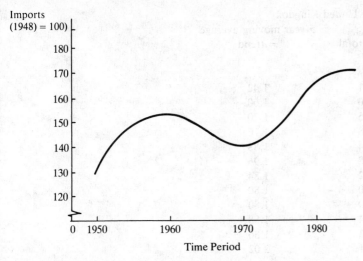

Imports
(1948) = 100)

180 —
170 —
160 —
150 —
140 —
130 —
120 —

0 1950   1960   1970   1980

Time Period

Figure 5.2 Trend showing cyclical influences

part on the time of year we are considering. Every housewife knows that the price of flowers rises as we approach Mother's Day, or that the price of tomatoes is higher in winter than it is in summer. You can multiply these examples indefinitely from your own experience, and anyone concerned with planning must take account of them.

(c) *Residual influences.* These are random external events which affect our variables. Sometimes the effect is negligible; at other times it is great; some occurrences will increase our figures; others will reduce them. We cannot see what is going to happen in the future, and so we cannot forecast such events. It is, however, reasonable to assume that in the long run they tend to cancel each other out, and that in our analysis we may initially ignore their impact.

(d) *Cyclical influences.* As the economy expands during a period of boom we would expect to find that such data as sales, output or consumer expenditure also show a rising trend; and during a period of slump we expect them to show a downward trend. Thus a wavelike motion may be observed in the pattern of our data. If we were to take a sufficiently long period, covering several cycles, we might even find a trend superimposed on the cyclical pattern – each cycle being generally higher, or lower, than the last.

Now, it is one thing to identify the factors which may cause the fluctuations in our data; it is quite another matter to disentangle one from the other and to measure its influence. We will try, in the next section, to isolate and measure the trend.

5.3   The calculation of trend

In Fig. 5.1 we have drawn a straight line through the middle of the series and said that the trend is rising 'something like this'. For serious statistical work, however, it is no use proceeding by guesswork. If we are going to use the trend figures we must calculate them, not guess them.

Now there are several methods of calculating trend, but the one we will explain first is a general all-purpose method – the method of *moving averages*. We will illustrate this by calculating the trend of unemployment in the United Kingdom during a 16-year period.

Unemployment in the United Kingdom

Year	Unemployment (%)	5-year total	5-year moving average = trend
1	1.3		
2	1.1		
3	1.2	7.1	1.42
4	1.4	8.0	1.60
5	2.1	8.5	1.70
6	2.2	8.8	1.76
7	1.6	9.4	1.88
8	1.5	9.8	1.96
9	2.0	9.2	1.84
10	2.5	9.0	1.80
11	1.6	9.0	1.80
12	1.4	9.4	1.88
13	1.5	9.3	1.86
14	2.4	10.1	2.02
15	2.4		
16	2.4		

Source: Employment & Productivity Gazette

A moving average is a simple arithmetic mean. We select a group of numbers at the start of the series, in this case the first five, and average them to obtain the first trend figure:

$$(1.3 + 1.1 + 1.2 + 1.4 + 2.1) \div 5 = 7.1 \div 5 = 1.42$$

This trend figure is placed opposite the centre of the group of five numbers, that is, opposite year 3.

To obtain the second trend figure we drop the first number of this initial group (1.3) and include the next number in the series (2.2). So the second group of five numbers is:

$$1.1 + 1.2 + 1.4 + 2.1 + 2.2 = 8.0$$

and the second trend figure is

$$8.0 \div 5 = 1.60$$

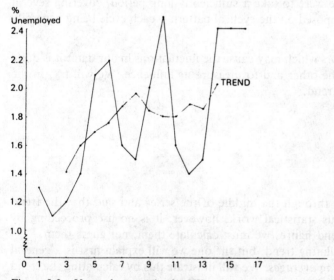

Figure 5.3   Unemployment in the UK

which is placed opposite year 4. We now drop the figure for year 2 (1.1) and include that for year 7 (1.6), giving a third trend figure of 1.70. We carry on in this way until all the data is exhausted.

You will notice that, using this method, there is no trend figure for the first two years nor for the last two. Do not forget this if you are asked to plot the original data and the trend on the same graph. Look carefully at Fig. 5.3 where the data has been plotted.

The trend is probably not as smooth as you thought it would be. True, the large-scale fluctuations have been eliminated, and up to year 8 it is fairly smooth, but from this point on there is no clear indication of which way the trend is moving. This was probably a result of changes in government policy as the United Kingdom struggled to counter inflation, prevent large-scale unemployment and promote economic expansion.

You are probably wondering why we chose a five-year moving average rather than a three-year or a seven-year one. There is no infallible rule you can follow. The correct time period to use is the one that gives the smoothest trend. A good working rule, however, is to look at the peaks and the troughs in the series. Troughs occur in years 2, 8 and 12; peaks occur in years 6, 10 and 14–16. It looks as if unemployment was subject to a five-year cycle and, in fact, economists confirm that this was so. Hence we chose a five-year moving average.

5.4 The trend of a quarterly series

A great deal of the data statisticians deal with is given quarterly or monthly, rather than annually. We saw in the previous section that one of the characteristics of such a series is that peaks usually occur in the same quarter of every year, as do the troughs; so the correct moving average to use is the four quarterly moving average. This does not demand any new concepts, but it does raise a practical problem. When we calculate the four quarterly moving average and it is placed opposite the mid-point of the group of numbers to which it refers, it falls *between* the second and third figures. Thus we cannot relate our figures of trend to any particular quarter. To see how we overcome this let us calculate the trend of the quarterly series of imports of raw material into Ruritania. Look at the table below.

Year	Quarter	Imports	Sum in 4's	Sum of two 4's	Trend
19-2	1	114			
	2	142			
			547		
	3	155		1,096	137.0
			549		
	4	136		1,106	138.25
			557		
19-3	1	116		1,112	139.0
			555		
	2	150		1,114	139.25
			559		
	3	153		1,130	141.25
			571		
	4	140		1,150	143.75
			579		
19-4	1	128		1,174	146.75
			595		
	2	158		1,209	151.125
			614		
	3	169		1,237	154.625
			623		
	4	159		1,268	158.5
			645		
19-5	1	137		1,313	164.125
			668		
	2	180		1,349	168.625
			681		
	3	192		1,370	171.25
			689		
	4	172		1,392	174.0
			703		
19-6	1	145			
	2	194			

Notice that when we add up in groups of four, the total is placed midway between the relevant quarters. In itself there is little wrong with this, but it prevents any comparison between the level of imports and the trend – a comparison we will have to make if we are to continue the analysis. To eliminate the problem, we use a technique known as *centreing*. At present, we have a trend figure of $547 \div 4 = 136.75$ placed opposite quarter $2\frac{1}{2}$, and one of 137.25 placed opposite quarter $3\frac{1}{2}$. If we take the average of these two it would be placed opposite quarter 3, and we have centred the trend so as to relate it to a specific quarter, i.e.

$$\frac{136.75 + 137.25}{2} = 137.0$$

The easiest way to do this in practice is to total successive pairs of four quarterly totals (e.g. $547 + 549$; $549 + 557$) and divide the resultant totals by 8 to give the trend. This is done in the last two columns of the table.

5.5 The calculation of seasonal variation

We have defined seasonal variation as an upswing and downswing in the value of the data. If we are to measure the magnitude of these fluctuations we must have a point of reference from which to measure. After all, when we say a mountain is 3,000 feet high, what we really mean is that it is 3,000 feet *above sea level*. It seems logical that we should measure the magnitude of the seasonal swing as the deviation from our calculated trend figure. So seasonal variation is not merely an upswing and downswing – it is a swing around the trend line. Armed with this definition, we can now calculate the quarterly variation in imports from the figures. The variation in any particular quarter will be:

(Original data *minus* calculated trend)

Year	Quarter	Imports	Trend	Deviation from trend
19-2	1	114		
	2	142		
	3	155	137.0	+ 18.0
	4	136	138.25	− 2.25
19-3	1	116	139.0	− 23.0
	2	150	139.25	+ 10.75
	3	153	141.25	+ 11.75
	4	140	143.75	− 3.75
19-4	1	128	146.75	− 18.75
	2	158	151.125	+ 6.875
	3	169	154.625	+ 14.375
	4	159	158.5	+ 0.5
19-5	1	137	164.125	− 27.125
	2	180	168.625	+ 11.375
	3	192	171.25	+ 20.75
	4	172	174.0	− 2.00
19-6	1	145		
	2	194		

The meaning of the deviations we have calculated in the last column is that in that particular quarter seasonal and other influences have caused imports to vary from trend by the calculated amount. Thus, in the third quarter of 19-2 these influences cause imports to rise 18.0 points above trend; in the third quarter of 19-3, 11.75 points above trend; in the third quarter of 19-4, 14.375 points above, and so on. Why should the deviations be so different in the same quarter of the year? The answer lies in the fact that seasonal and *other* influences have been at work. The deviations have been caused by both seasonal and residual influences.

While we cannot separate these two influences, it is reasonable to believe that while seasonal influences will always operate in the same direction, residual influences will sometimes raise the figures and at other times will lower them. If, then, we take a sufficiently long series and take the *average* deviation for any particular quarter, the residual influences will tend to offset each other and we will be left with the purely seasonal variation. This is the rationale behind the calculation of seasonal variation, and the reason why it is often called *average seasonal variation*.

We will now pick up the quarterly deviations from trend and tabulate them in order to calculate seasonal variation (SV).

Year	Quarter			
	1	2	3	4
19-2			+ 18.0	− 2.25
19-3	− 23.0	+ 10.75	+ 11.75	− 3.75
19-4	− 18.75	+ 6.875	+ 14.375	+ 0.5
19-5	− 27.125	+ 11.375	+ 20.75	− 2.0
Total	− 68.875	+ 29.0	+ 64.875	− 7.5
Average	− 22.958	+ 9.667	+ 16.219	− 1.875 = + 1.053
Adjust	− 0.263	− 0.263	− 0.263	− 0.263
	− 23.221	+ 9.404	+ 15.956	− 2.138
SV	− 23	+ 9	+ 16	− 2

There are several things to note about this table. First, you will see that there are four deviation figures for the third and fourth quarters, but only three for the first and second. This is common and depends simply on the lengths of the series we are examining, but we must not forget it when we are calculating the average quarterly deviation. Secondly, the total of the average deviations should be zero (remember the definition of the arithmetic mean). In fact, the total is + 1.053 and, since we know it should be zero, an adjustment is necessary. We do not know where the difference springs from, so we adjust each quarterly average by the same amount, 0.263 (i.e. 1.053 ÷ 4). Since the total deviations are too many, we will have to subtract 0.263 from each quarterly average. Thirdly, when we have made all these adjustments, we are left with average deviations correct to three places of decimals. Now this is silly! In the course of our calculations we have made a number of assumptions. Each one is a logical, reasonable assumption, but we must not pretend to a degree of accuracy we cannot guarantee. Since the original data is given in integers only, it is better to give average seasonal variation in integers also. So we will quote the seasonal variation as − 23, + 9, + 16, − 2. Notice that the total still comes to zero.

What do these figures mean? Using the figures for the first quarter as an example, it merely says that economic conditions are such that the index of imports will tend to fall below the trend, and that on average it will be 23 points below. On the other hand, in the second quarter of the year, the economic climate is different, imports tend to rise above trend and on average are 9 points above it.

5.6 Series with seasonal variation eliminated

Useful as a knowledge of seasonal variation may be to a firm planning for the future, there is little doubt that the constant fluctuations tend to hide the underlying behaviour of the variable.

Comparisons and assessments of performance over time are difficult to make. When a particular salesman returns with a full order book, is it because he has made a superhuman effort or merely because the seasonal swing is in his favour? The need to make assessments of this nature has resulted more and more in data being produced 'seasonally adjusted' or 'with seasonal variation eliminated'.

Adjusting data to eliminate seasonal variation is a relatively simple matter. If seasonal variation is $+23$, we are, in effect, saying that our data is 23 points above the trend because of seasonal influences, and we must, therefore, reduce our figures by 23 to eliminate these influences. Similarly, if seasonal variation is -23, we must add 23 to our figures. If we remember that $(--) = (+)$, we may formulate the rule that

series with seasonal variation eliminated = original data − seasonal variation

Applying this rule to the data, we are analysing:

Year	Quarter	Import index	Seasonal variation	Series with seasonal variation eliminated
19-2	1	114	-23	137
	2	142	$+9$	133
	3	155	$+16$	139
	4	136	$+2$	138
19-3	1	116	-23	139
	2	150	$+9$	141
	3	153	$+16$	137
	4	140	-2	142
19-4	1	128	-23	151
	2	158	$+9$	149
	3	169	$+16$	153
	4	159	-2	161
19-5	1	137	-23	160
	2	180	$+9$	171
	3	192	$+16$	176
	4	172	-2	174
19-6	1	145	-23	168
	2	194	$+9$	185

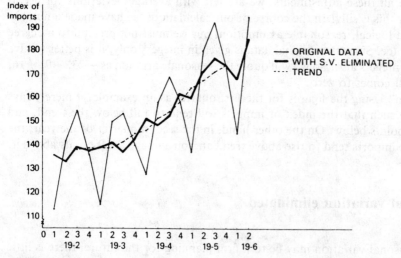

Figure 5.4

We now have three different series obtained from the same data: the original figures, the trend and the series with seasonal variation eliminated. When you eliminate the seasonal variation, you might, of course, expect to be left with a series approximating to trend. Do not forget, however, that residual influences will still be found in this series, and the more important the residual influence, the more the series will differ from trend. In Fig. 5.4 we have plotted the three series on the same graph. What conclusions can we draw? First, the trend shows us that there has been a steady and unbroken rise in the volume of imports over the whole period, which shows little signs of ceasing. This may be a result of an expanding industry requiring more and more raw materials, or it may be a result of the domestic consumer buying more and more foreign consumer goods. The graph will not tell us why something happens, but it will lead us to ask the questions.

Secondly, the trend and the series with seasonal variation eliminated tend to run closely together. There are differences, but they are small. So we can come to the conclusion that residual influences are fairly unimportant and do not drastically affect the level of imports. This may seem strange, but in fact events such as a dock strike which largely cuts off the flow of goods are very rare.

Nor do such residual influences as there are always work in one direction. Sometimes they raise the level of imports above trend; at other times they lower them; and the pattern is regular. In fact, so regular is the pattern that one can speculate that residuals here have a double effect. At first they slow down the level of imports, but fairly quickly a back-log of orders builds up and, as a result, in the next quarter imports have to rise as the back-log is cleared.

5.7   The importance of residuals

You can find out a great deal about residual influences from a graph such as Fig. 5.4, but it does not tell us why they are important.

All planning, whether financial or not, is based on forecasting, and if the forecast cannot be relied on planning is a waste of time. Now, the characteristic of a residual is that it cannot be included

Year	Quarter	Imports	Trend	Seasonal variation	Residual Absolute	Residual As percentage of imports
19-2	1	114		−23		
	2	142		+9		
	3	155	137.0	+16	+2	1.3
	4	136	138.25	−2	−0.25	0.18
19-3	1	116	139.0	−23	0.0	−
	2	150	139.25	+9	+1.75	1.17
	3	153	141.25	+16	−4.25	2.78
	4	140	143.75	−2	−1.75	1.25
19-4	1	128	146.75	−23	+4.25	3.32
	2	158	151.125	+9	−2.125	1.34
	3	169	154.625	+16	−1.625	0.96
	4	159	158.5	−2	+2.5	1.57
19-5	1	137	164.125	−23	−4.125	3.01
	2	180	168.625	+9	+2.375	1.32
	3	192	171.25	+16	+4.75	2.47
	4	172	174.0	−2	0.0	−
19-6	1	145		−23		
	2	194		+9		

in the forecast so, when its influence is felt, the forecast is upset. In analysing our series of imports we have already discovered that the influence of residuals is fairly small. So, it is probable that though our forecast may be upset, we will not be very far wrong. On the other hand, if the residual influence is great, our forecasts may be badly upset.

Before we come to any decisions, however, we must obtain a quantitative measure of the extent of residual variations. We know that the original data is composed of

trend + seasonal variations + residual variations

and it follows then that

residuals = original data − trend − seasonal variations

The table on the previous page brings together the calculations we have made so far. Our calculations confirm what we suspected from the graph. The residual influence never affects our figures by more than 4.75 points and, what is more important, on only two occasions is there more than a 3 per cent influence. Thus, if our forecast is otherwise accurate, we would expect it to be correct to within 3 per cent in spite of the residual influences.

5.8 Forecasting from the time series

The key words in the last paragraph are 'if our forecast is otherwise accurate'. Naturally, in looking to the future, absolute accuracy is difficult to attain but, nevertheless, very good forecasts can be made.

The basis of the forecasting we will do lies in our knowledge of the behaviour of trend. The trend will not suddenly change direction. As you can see, in Fig. 5.5 the trend is beginning to rise rather more slowly. Over the next few quarters the rate of rise may continue to fall and eventually the trend will reach a maximum and perhaps begin to fall. Or it may slowly change direction upwards with an accelerated rate of rise. The point is that it is not going to change *suddenly*. So, we can extend (or project) the trend forward over the next two or three quarters, confident that our projection will be fairly accurate. The further we project the trend, of course, the greater is the possibility of inaccuracy, and great care is needed in making the projection.

In Fig. 5.5 the trend is drawn and a suggested projection made, extending over the next two quarters. We make no claim, of course, that this projection is perfect, and you may make a far more accurate one yourself. Our new trend values are:

19-6 Quarter 1 176.5
2 178.75

To understand just how we can use these figures, we will go back to the calculation of trend and reproduce the last part of the calculations.

Year	Quarter	Imports col. 1	Sum in 4's col. 2	Sum of two 4's col. 3	Trend col. 4
19-5	3	192	681	1,370	171.25
	4	172	689	1,392	174.0
19-6	1	145	703	$(8x = 1,412)$	$(x = 176.5)$
	2	194	$(8x - 703 = 709)$	$8y =$	$(y = 178.75)$
	3	$(a = 198)$			
		$b =$			

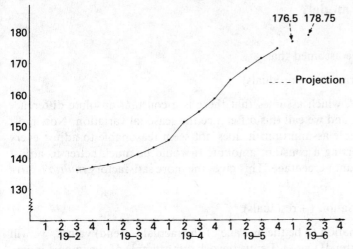

Figure 5.5

We will call the forecast values for 19-6, quarter 3, *a* and that for quarter 4, *b*. Suppose the trend value we read from the graph is *x*, which is placed in column 4. But we obtain *x* by dividing column 3 by 8, so in column 3 we must place $8x$. In its turn this $8x$ is the sum of 703 plus an unknown number, which will have therefore a value $8x - 703$, which is the next figure in column 2. Column 2 is the sum of four import indices and $8x - 703 = 145 + 172 + 194 + a$. It follows then that:

$$a = 8x - 703 - 172 - 145 - 194 = 8x - 1,214$$

Since $x = 176.5$,

$$a = (8 \times 176.5) - 1,214 = 1,412 - 1,214 = 198$$

This seems to be a reasonable forecast since the third quarter is always slightly higher than the second quarter (which was 194). The values we have calculated have been placed in the appropriate columns in the table. Now, using these figures, and given that the value of the trend *y* is 178.75, calculate for yourself the value of the import index for the fourth quarter of 19-6.† The answer is given in the footnote, and you will probably agree that the two figures we have forecast are entirely consistent with the pattern of the series.

Forecasting by projecting the trend of an annual series is exactly the same as for a quarterly series. Suppose we are taking a five-year moving average, and the last few rows of the calculations are:

Year	Output	5-year total	Trend
19-2	15	70	14.0
19-3	18	74	14.8
19-4	13	76	15.2
19-5	14	?	?
19-6	16		
19-7	?		

The first step in forecasting a value for 19-7 would be to project the trend forward. We could then read a value for 19-5. Suppose this value of the projected trend were 15.6. This figure is the average of a five-year total in the previous column, so that total must be $5 \times 15.6 = 78$. This in itself is the sum of $18 + 13 + 14 + 16 + $ (output in 19-7). Thus output in 19-7 would be forecast as

$$78 - 18 - 13 - 14 - 16 = 17$$

† Answer: $y = 178.75$ $b = 8y - 709 - 145 - 194 - 198 = 8y - 1,246 = 1,428 - 1,246 = 182$.

5.9 Additive or multiplicative model?

In our analysis of time series, we have assumed that

actual data = trend + seasonal variation (+ residuals)

This is the so-called *additive model*, which assumes that there is a constant absolute difference between the actual data and the trend, and we call this difference the seasonal variation. Now most statisticians think this is not a reasonable assumption: it does not seem reasonable to adjust every (say) first quarter by adding or subtracting a constant amount. It would be much better to adjust every corresponding quarter by a constant percentage. This gives the more satisfactory *multiplicative model*:

actual data = trend × seasonal variation (+ residuals)

Instead, then, of taking seasonal variation to be the average of the (actual − trend) values, we will take it to be the average of the (actual/trend) values. Let us rework our example of the import index, this time using the multiplicative model. The trend is calculated in exactly the same manner as for the additive model − by using a four quarter, centred, moving average.

Year	Quarter	Import index	Trend	Actual Trend
19-2	1	114		
	2	142		
	3	155	137.000	1.1313
	4	136	138.250	0.9837
19-3	1	116	139.000	0.8345
	2	150	139.250	1.0771
	3	153	141.250	1.0831
	4	140	143.750	0.9739
19-4	1	128	146.750	0.8722
	2	158	151.125	1.0454
	3	169	154.625	1.0929
	4	159	158.500	1.0031
19-5	1	137	164.125	0.8347
	2	180	168.625	1.0674
	3	192	171.250	1.1211
	4	172	174.000	0.9885
19-6	1	145		
	2	194		

	Quarter			
Year	1	2	3	4
19-2			1.1313	0.9837
19-3	0.8345	1.0771	1.0831	0.9739
19-4	0.8722	1.0454	1.0929	1.0031
19-5	0.8347	1.0674	1.1211	0.9885
Total	2.5414	3.1899	4.4284	3.9492
Average	0.8471	1.0633	1.1071	0.9873

We would predict the total of the average ratios to be 4, whereas it is 4.0048. So we must adjust the averages by multiplying each average ratio by 4/4.0048 to give seasonal variation ratios.

	Quarter		
1	2	3	4
0.846	1.062	1.106	0.986

If we wish to find a deseasonalised series, we would divide the actual data by the seasonal variation.

Year	Quarter	Index	Seasonal ratio	Deseasonalised Index
19-2	1	114	0.846	135
	2	142	1.062	134
	3	155	1.106	140
	4	136	0.986	138
19-3	1	116	0.846	137
	2	150	1.062	141
	3	153	1.106	138
	4	140	0.986	142
19-4	1	128	0.846	151
	2	158	1.062	149
	3	169	1.106	153
	4	159	0.986	161
19-5	1	137	0.846	162
	2	180	1.062	169
	3	192	1.106	174
	4	172	0.986	174
19-6	1	145	0.846	171
	2	194	1.062	183

Review questions

5.1   Cost of raw materials

Year	(£ per ton)
1	3.38
2	4.11
3	4.17
4	3.87
5	3.26
6	3.81
7	3.89
8	4.45
9	4.30
10	3.65
11	5.00

Construct a five-year moving average trend line.

5.2   A traffic census taken at the same time each day during July on a busy road approaching a holiday resort revealed that the number of vehicles passing per hour was as follows (to the nearest 10):

	Week				
	1	2	3	4	5
Monday		840	840	830	820
Tuesday		860	860	830	
Wednesday		1,190	1,200	1,220	
Thursday		840	830	840	
Friday		970	1,020	1,080	
Saturday	1,800	1,860	1,950	2,100	
Sunday	1,460	1,480	1,520	1,550	

(a) Plot the data on a graph.

(b) Calculate a suitable moving average and plot the trend line.

5.3 Number of visitors to a certain hotel

	Quarters			
	1	2	3	4
19-2	58	85	97	73
19-3	64	96	107	89
19-4	76	102	115	94

(a) Draw a graph of this data.

(b) Calculate a four quarter centred moving average trend line, and mark this on your graph.

Keep your answers to this question safe as you will need them for the next question.

5.4 Using your workings for the previous question answer the following.

(a) Calculate the average seasonal variation using the multiplicative method.

(b) Find the seasonally adjusted series and plot this on your graph.

(c) Look carefully at your graph. In 19-4, quarter 2, the number of visitors was up on the previous quarter though, seasonally adjusted, the number of visitors fell. How can you account for this?

(d) Suppose that the estimated trend figure for 19-4, quarter 4, was 98. Predict the number of visitors for that quarter.

6 Regression and correlation

6.1 Introduction

In this chapter, we are concerned with examining two sets of data that we suspect are connected. We will attempt to discover the nature of this connection so that, given one value, we can predict its associated value (for example, given output we should be able to predict the associated level of cost). Such analysis is called regression, and you will also be shown how regression analysis can be applied to a time series. Finally, we attempt to measure the strength of the connection using correlation analysis.

If you have listened carefully to news bulletins or read the business pages of a responsible newspaper, you will probably have come across the phrase 'economic indicator' or 'business indicator'. Such an indicator is an event which, although it may be important in its own right, is of even more importance in helping to predict what is going to happen to some other variable in the future. It is commonly believed, for example, that government expenditure will affect the level of employment. What is important, of course, is knowledge of how much extra expenditure is needed to create, say, 1,000 new jobs, and what is the time lag before the unemployed actually start work. Surely a person who can identify and quantify relationships such as this is worth their weight in gold to any company.

No book can teach you to identify every relationship which may be important in a particular set of circumstances, but you should have a good idea of those relationships which are relevant to your own work. You will know, for example, that next month's sales figures are influenced by this month's advertising, or that the major influence on your departmental budget is the behaviour of labour costs. What you have to learn is how a knowledge of such relationships can be used to enable you to make predictions (or estimates). After all, in forecasting sales figures or preparing a departmental budget you *are* being asked to make predictions.

6.2 Bivariate distributions

At some time, every businessman must have faced the problem as to whether or not he should advertise. Some believe that their sales depend on the amount of advertising they undertake; others hold equally firmly to the view that the key to increasing their sales lies in providing service, in offering a wide selection of goods, or in giving an unconditional guarantee of satisfaction. The only way of settling this sort of argument for any individual firm is to examine the facts.

Brown's Department Store has recently hired a consultant to help boost its sales and thereby the morale of its management (and hopefully its employees). The consultant, a marketing specialist, has reasoned that sales are a function of advertising expenditure. Thus he has decided to try to relate sales to the amount spent on advertising. The data from the company's records are produced for him by the accountant and he is presented with the information in the table below.

	Advertising X (£000)	Sales Y (£m)
1976	100	9
1977	105	8
1978	90	5
1979	80	2
1980	80	4
1981	85	6
1982	87	4
1983	92	7
1984	90	6
1985	95	7
1986	93	5
1987	85	5
1988	85	4
1989	70	3
1990	85	3

Now this series is rather different from others we have considered. Whereas in the series we have so far studied each item has had a single value, in this series there are two items associated with each value. We call distributions like this *bivariate distributions*.

We cannot, of course, draw any meaningful conclusions from the data as it stands. Possibly it would help if we were to present the information in the form of a diagram – in this case the scatter diagram.

Before we get too involved with the scatter diagram itself, let us consider the variables which are the subject of analysis. First, what is the nature of the data which comprises the two variables? Can it be treated as an enumeration of the population, or must we treat it as a sample? The answer may seem very simple, but there are serious implications underlying it. We shall be calculating measures which contain constants. If we treat the data as a population, we will treat the constants we calculate, α and β (alpha and beta), as population parameters, absolutely accurate and unchanging. If the data, however, is treated as a sample, the constants we calculate will be a and b, which are no more than estimates of α and β. As such they might be wrong and will certainly be subject to some sampling error.

Secondly, which is the dependent and which the independent variable? In terms of the analysis, we shall assume one variable to be a function of the other. But in terms of the actual data, the analysis itself cannot determine which is which. This determination can be made only on the basis of your own knowledge of the subject matter. In so far as Brown's Department Store is concerned, there is little doubt that most people would agree that advertising expenditure is the independent variable and sales the dependent. We shall proceed to analyse on this assumption, even if one or two of you do disagree.

6.2.1 The scatter diagram

The purpose of the scatter diagram is to illustrate diagrammatically any relationship that may exist

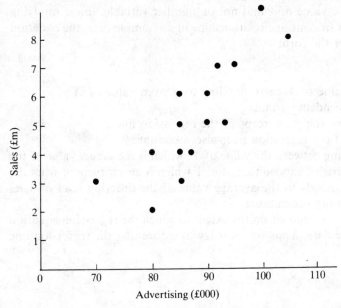

Figure 6.1   Brown's Department Store, sales and advertising 1966–80

between the dependent and independent variables. To the extent that it succeeds it can help the analyst in three ways:

(a) It indicates generally whether or not there appears to be a relationship between the two variables.

(b) If there is a relationship it may indicate whether it is linear or non-linear.

(c) If the relationship is linear, the scatter diagram will show whether it is negative or positive.

If you look at Fig. 6.1 you will see that the data does provide some evidence that the higher the level of advertising, the higher the level of sales achieved. Certainly the same level of advertising is associated occasionally with differing levels of sales, just as the same level of sales is associated with differing levels of advertising. But as you can see, the general trend of the points on this diagram is an upward slope from left to right. It is indicated that there is a relationship, that it is linear and that it is positive in the sense that as one variable rises so does the other.

The scatter diagram, then, is the graph of a bivariate distribution and is used to obtain an indication of whether association exists between the two variables. Having discovered that in this case an association is indicated, we shall now try to discover the nature of the relationship and the degree of association between the variables. If we can do that we can use our knowledge to estimate what the value of sales will be for any given level of advertising.

6.3   Regression

First we will attempt to discover the nature of the relationship by calculating an equation enabling us to estimate the value of one variable given that we know the value of the other. The variable we are trying to predict is the dependent variable (the one plotted conventionally on the *y* axis) while the variable we are using as a basis for prediction is the independent variable (plotted on the *x* axis). Initially we will confine ourselves to an analysis of the simple linear case, simple implying two variables only such that y depends on the value of x and not on another variable, linear implying that the relationship between x and y is a straight line relationship. In this simple case, the equation which best fits the data is an equation of the form

$$\hat{Y} = a + bX$$

where $\hat{Y}$ is an estimate of the average value of Y corresponding to a given value of X

X is the actual value of the independent variable

a is a constant − an estimate of α, the y intercept of the regression line

b is an estimate of β, the slope of the regression line; also a constant

You will notice that we are distinguishing between the value of Y, which is the *actual* value of the sales resulting from a given level of advertising expenditure, and $\hat{Y}$ which is an *estimate* of what the sales will be. Our estimate in fact corresponds to the average value of the differing level of sales resulting from the same value of advertising expenditure.

The accuracy of an estimate of this nature depends on the extent to which the regression equation $\hat{Y} = a + bX$ and its graph actually fits the data. Thus we must try to ensure that the regression line is the line of best fit.

6.3.1 The regression line

One way of obtaining the line of best fit would be to place a ruler across the scatter diagram and draw in the line which 'looks right'. We could then find where this line cuts the y axis, to give us the value of the constant a, and calculate the gradient of the line, to give us the value of the constant b. But guessing the line of best fit is just not good enough, especially if decisions are to be taken on the basis of our estimates. What we need is some criterion by which we can determine what we mean by 'best fit'.

One criterion we could adopt would be to minimise the magnitude of the error involved, that is, the differences between the value of the points on the diagram (Y) and the value of the points on the regression line we have drawn ($\hat{Y}$). We could, that is, minimise the value of $\Sigma (Y - \hat{Y})$. Problems arise with this criterion, however. In Fig. 6.2(a) three errors are shown, one positive and two negative. They offset each other and the net error is zero. In this case, by minimising the error we have obtained what seems to be quite a good fit. Look now at Fig. 6.2(b).

The regression line, too, reduces the error to zero, yet quite clearly in this case the fit is bad.

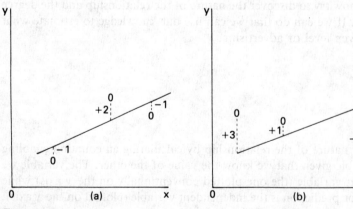

Figure 6.2

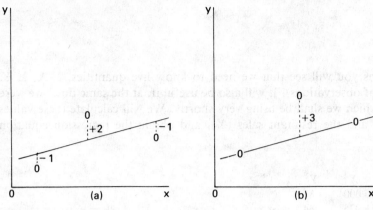

Figure 6.3

Reducing the error to a minimum, then, is no criterion for positioning the line of best fit. It may give a good fit, but it is just as likely to give a bad one.

What, then, if we were to minimise the absolute error, ignoring the sign? Once again this will not guarantee a line of best fit. In Fig. 6.3, the line in diagram (a) is clearly a better fit than the line in diagram (b), yet line (b) minimises the absolute error. Statisticians reject both of these criteria as unsatisfactory and have defined the line of best fit as 'that line which minimises the sum of the squares of the errors – the least squares regression line'.

The use of this criterion has several advantages. First, by squaring the errors we remove all negative signs and so avoid negative errors cancelling out the positive ones; second, squaring tends to emphasise the larger errors and the minimising criterion means that we are trying to avoid these; third, all points are taken into account so the criterion avoids lines such as that in Fig. 6.3(b).

6.3.2   The least squares regression line

How can we fit a line to our data which minimises the sum of the squares of the errors? We know that the equation of the line we need is

$$\hat{Y} = a + bX$$

The statistics necessary to find a and b are the variance of X, and what we call the covariance of X and Y. The variance of X we will symbolise as s_x^2, and as you know this is calculated as:

$$s_x^2 = \frac{\sum x^2}{n} - \left(\frac{\sum x}{n}\right)^2$$

The covariance is symbolised as s_{xy}^2 and it is calculated as:

$$s_{xy}^2 = \frac{\sum XY}{n} - \frac{\sum X \sum Y}{n \cdot n}$$

The expression from which you can calculate the value of b is:

$$b = \frac{\text{covariance}}{\text{variance}} = \frac{s_{xy}^2}{s_x^2}$$

and the expression for a is

$$a = \frac{1}{n} (\Sigma Y - b \Sigma X)$$

If you examine these equations you will see that we need to know five quantities, ΣX, ΣY, ΣXY, ΣX^2 and n (the number of observations). It will also be useful if, at the same time, we were to calculate the value of ΣY^2, which we shall be using very shortly. We will calculate these values for advertising expenditure (X) and the resultant sales (Y), and obtain the regression equation connecting the two.

X (£000)	Y (£m)	XY	X^2	Y^2
100	9	900	10,000	81
105	8	840	11,025	64
90	5	450	8,100	25
80	2	160	6,400	4
80	4	320	6,400	16
85	6	510	7,225	36
87	4	348	7,569	16
92	7	644	8,464	49
90	6	540	8,100	36
95	7	665	9,025	49
93	5	465	8,649	25
85	5	425	7,225	25
85	4	340	7,225	16
70	3	210	4,900	9
85	3	255	7,225	9
1,322	78	7,072	117,532	460

Substituting, we find:

$$s_{xy}^2 = \frac{\Sigma XY}{n} - \frac{\Sigma X \, \Sigma Y}{n \cdot n} = \frac{7,072}{15} - \frac{1,322 \times 78}{15 \times 15} = 13.17333$$

$$s_y^2 = \frac{\Sigma X^2}{n} - \left(\frac{\Sigma X}{n}\right)^2 = \frac{117,532}{15} - \left(\frac{1,322}{15}\right)^2 = 67.98222$$

Hence $b = \dfrac{13.17333}{67.98222} = 0.193776$

and $a = \dfrac{1}{15} (78 - 1,322 \times 0.193776) = -11.8781$

So the regression equation linking advertising expenditure and sales is:

$$\hat{Y} = -11.8781 + 0.1938X$$

How well does this equation fit the original data? Let us calculate what our estimate of the value of Y would be for the various values of X, and then compare them with the observed values. Using the regression equation,

$$\hat{Y} = -11.8781 + 0.1938X$$

we find that:

For a value of X	Observed Y	Estimated $\hat{Y}$
100	9	7.5019
105	8	8.4709
90	5	5.5639
80	2	3.6259
80	4	3.6259
85	6	4.5949
87	4	4.9825
92	7	5.9515
90	6	5.5639
95	7	6.5329
93	5	6.1453
85	5	4.5949
85	4	4.5949
70	3	1.6879
85	3	4.5949

As you can see, the estimation of the value of Y is not perfect but, on the whole, we think you will agree, it is pretty good.

The equation we have calculated is known as the equation of the regression line of y on x because we are estimating the value of Y given the value of X. Thus if we had spent £75,000 on advertising we would estimate a sales figure by putting $X = 75$ (thousand) in the regression equation, i.e.

$$\hat{Y} = -11.8781 + (75 \times 0.1938) = 2.6569 = £2,656,900$$

What does this prediction mean? Obviously it cannot mean that every time we spend £75,000 on advertising we would have sales of exactly £2,656,900. Conditions could never be quite the same on each occasion, and the fact that we spent some money last month would, of itself, influence the effect of spending more this month, since advertising has a cumulative effect. In fact, actual sales would fluctuate, even if advertising expenditure did not. The figure we estimate for sales is, in fact, an average figure.

We calculated the regression equation from observed values of X ranging from £70,000 to £105,000 and used that equation to estimate the level of sales resulting from advertising expenditure of £75,000. When we take a value of the independent variable within the range of the observed values, we call the process interpolation (placing between). We might have asked, however, for the estimated value of sales from advertising expenditure of £120,000.

$$\hat{Y} = -11.8781 + (0.1938 \times 120) = £1,137,790$$

Here we used a value of X outside the range of observed values. We call this process extrapolation (placing outside). Both estimates are subject to error, but an extrapolated estimate is less reliable than an interpolated one. Can you see why? Within the observed range of X we know how the data behaves, and how well the straight line fits. Outside the observed range we do not know how the data behaves, and the straight line may no longer be a good fit for these values of X. This is illustrated in Fig. 6.4 and we can see that great caution is needed when making extrapolated estimates.

6.3.3 Linear regression and the time series

We saw in Chapter 5 that the trend of many time series is linear and we learned to calculate that trend using a moving average. Now although a time series is not strictly a bivariate series, we can obtain the linear trend by using regression analysis.

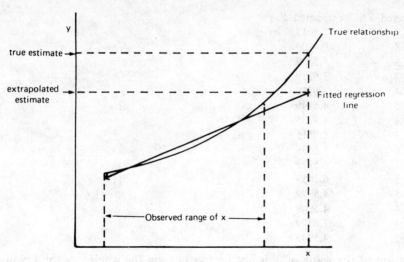

true estimate

extrapolated
estimate

True relationship

Fitted regression
line

Observed range of x

Figure 6.4

Suppose we are examining the figures of quarterly profit made by a firm. If the data appears to demand a linear trend, we can assume that profits depend on time and convert time into an independent variable by numbering successive quarters 1, 2, 3, 4, 5 and so on.

Profits have varied over the last four years quarter by quarter, but have been generally rising. The figures, in thousands of pounds are:

Year	Quarter			
	1	2	3	4
19-5	40	60	90	70
19-6	50	70	110	80
19-7	70	80	120	90
19-8	80	100	130	100

If we number the quarters from 1 to 16 and use this as the independent variable, we can convert the time series into a bivariate distribution.

Year	Quarter	X	Y	X^2	XY
19-5	1	1	40	1	40
	2	2	60	4	120
	3	3	90	9	270
	4	4	70	16	280
19-6	1	5	50	25	250
	2	6	70	36	420
	3	7	110	49	770
	4	8	80	64	640
19-7	1	9	70	81	630
	2	10	80	100	800
	3	11	120	121	1,320
	4	12	90	144	1,080
19-8	1	13	80	169	1,040
	2	14	100	196	1,400
	3	15	130	225	1,950
	4	16	100	256	1,600
		136	1,340	1,496	12,610

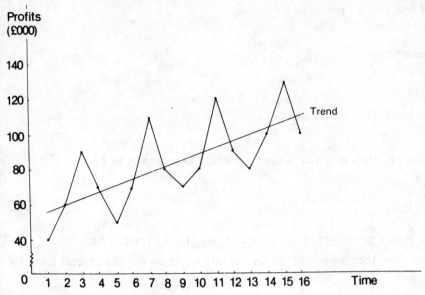

Profits
(£000)

Figure 6.5

$$b = \frac{12{,}610 - \dfrac{136 - 1340}{16}}{1{,}496 - \dfrac{(136)^2}{16}} = 3.59$$

$$a = \frac{1}{16}\,(1{,}340 - 136 \times 3.59) = 53.235$$

Thus the equation of our trend line for this series is:

$$\hat{Y} = 53.235 + 3.59X$$

We can obtain trend values by inserting into this equation $X = 1, 2, 3, \ldots, 16$. Having thus obtained the trend we could use it to calculate the deviations from it and to calculate the average seasonal variation. The original data and the trend are plotted in Fig. 6.5. As you know, only two points are needed to plot a straight line, and we have chosen $X = 1$ and $X = 10$.

when $X = 1$, $\hat{Y} = 56.825$ and when $X = 10$, $\hat{Y} = 89.135$

6.4   Curve fitting

We have concerned ourselves so far with relations that are linear. However, there are times when a curve would obviously be a better fit. Now there are many possible curves that could be investigated, and in this section we will see how to fit a curve in the form

$$y = ab^x$$

Consider the case of a firm whose sales in the current year are £1,100,000. It is anticipated that sales will grow at the rate of 5 per cent per year. For convenience, we will work in thousands, and we would predict sales in one year's time to be

$$1,100(1.05) = 1,115$$

and sales in two years' time to be

$$1,115(1.05)$$
$$= 1,100(1.05)^2 = 1,212$$

Likewise, sales in three years' time would be

$$1,100(1.05)^3 = 1,273$$

So, if y is sales in thousands, then in x year's time we would expect sales to be

$$y = 1,100(1.05)^x,$$

and sales would grow like this:

Year	1	2	3	4	5	6	7	8	9	10
Sales (£000)	1,155	1,212	1,273	1,337	1,404	1,474	1,548	1,625	1,706	1,792

If we graph sales against time, then we would obtain a gently rising curve. The general form for this curve is

$$y = ab^x$$

where y is sales and x is the year. In the case of our firm we know that $a = 1,100$ and $b = 1.05$. But suppose we did not know the value of a and b and were presented with a similar situation to the one above. How would we evaluate a and b?

Now according to the theory of logarithms

$$\log(ab) = \log a + \log b$$

and

$$\log(b^x) = x \log b$$

So using logarithms, we can say that the equation

$$y = ab^x$$

can be written in logarithms as

$$\log y = \log a + x \log b$$

Notice that this relationship is linear, so we can use the method of linear regression to obtain the logarithms of a and b. The difference is that instead of using actual sales, we must use the logarithm of sales. Our calculations would proceed something like this:

Year x	log sales y	x^2	xy
1	3.0626	1	3.0626
2	3.0832	4	6.1664
3	3.1048	9	9.3144
4	3.1262	16	12.5048
5	3.1473	25	15.7365
6	3.1685	36	19.0110
7	3.1897	49	22.3279
8	3.2108	64	25.6864
9	3.2319	81	29.0871
10	3.2534	100	32.5340
55	31.5784	385	175.4311

$$\log b = \frac{175.4311 - \dfrac{55 \times 31.5784}{10}}{385 - \dfrac{55^2}{10}} = \frac{1.7499}{82.5} = 0.0212$$

$\log a = 1/10[31.5784 - 0.0212 \times 55] = 3.0412$
If $\log a = 3.0412$, then $a = 1{,}100$
If $\log b = 0.0212$, then $b = 1.05$

which agrees exactly with our known values for a and b.

So far in this chapter we have been concentrating on the nature of the relationship between two variables but we must also pay attention to the degree of relationship. We will now turn our attention to measuring the extent to which two variables are related when that relationship is less than perfect.

6.5 Correlation

Our analysis of regression has shown us that it is possible to obtain an expression showing the relationship between two variables such that for any value of X it is possible for us to estimate the value of Y associated with it. Sometimes the estimate we make is consistently good and there seems to be a very close relationship between the two variables. At other times the estimate is very poor indeed and we are led to think that the relationship is weak. Naturally we would like to know more about this relationship and in particular we would like to be able to measure its strength. Statisticians have designed a measure, the coefficient of correlation, which measures the relationship between the variables by asking the question, 'how much of the variation in the value of Y can be explained by variations in the value of X?'

Suppose we take an extreme case. Having calculated the regression equation we find that the regression line is such that every point in the scatter diagram falls exactly on the line. The whole of the variation in Y is explained by variations in the value of X. We have here a case of what we call perfect correlation. Now this definition has one great drawback. Figure 6.6 illustrates two cases of perfect correlation, though we can see that they are of quite different types. In case (a) we notice that y increases as x increases, and call this a *positive correlation*. In case (b), however, y decreases as x increases and we call this a *negative correlation*. It would seem sensible, then, not merely to measure correlation, but also to state whether it is positive or negative.

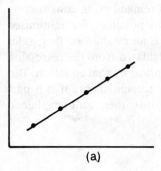
(a)

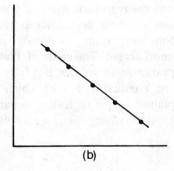
(b)

Figure 6.6

There is a second problem too, though it is not quite so obvious. We always tend to assume that if the correlation between two variables is high that variations in one are caused by variations in the other. Now association has no causal implications. There is a very high association between admissions into mental institutions and the number of TV licences issued. But it would be a bold man who maintained that we have proved that watching television causes insanity. We have here a case of spurious or nonsense correlation and we must always beware of this. In using the correlation coefficient we must be sure in our own mind that association between two variables is likely or, at least, possible.

Bearing in mind this warning, how will we go about measuring correlation?

6.5.1 Concepts of variation

So important is the idea of variation in the analysis of correlation that it would be as well to look at it much more closely. First we will draw a scatter diagram of our data and superimpose on it a line representing the mean value of Y ($\bar{Y}$) (Fig. 6.7(a)). This line will, of course be a horizontal straight line since the mean value of Y is the same for all values of X. As you can see, there is a great deal of variation of the values of Y about $\bar{Y}$, and we will measure the total variation by totalling the squares of the individual deviations. Can you see why we must square the deviations? Obviously if we have calculated the value of $\bar{Y}$ correctly, the sum of the deviations must be equal to zero. Thus, as a measure of total variation, we have:

sum of the squares of the deviations from $Y = \Sigma(Y - \bar{Y})^2$

You will have realised already that if we take the average of the deviations squared we will have the variance of Y.

$$s_y^2 = \frac{\Sigma(Y - \bar{Y})^2}{n} = \frac{\Sigma Y^2}{n} - \left(\frac{\Sigma Y}{n}\right)^2$$

If you turn now to Fig. 6.7(b), here we have inserted on the scatter diagram not $\bar{Y}$, but the regression line. Here we have quite a different concept of variation. Each point on the regression line differs from the value of $\bar{Y}$, and we could consider this variation ($\hat{Y} - \bar{Y}$). Again we will take the sum of the squares of the deviations, $\Sigma(\hat{Y} - \bar{Y})^2$. Now there is an important characteristic of this variation. The value of $\hat{Y}$ is in each case calculated from the regression equation and hence is determined by the value given to X. Thus we can say that this variation is entirely explained by the regression equation.

Finally, you will notice that there is yet another variation we might consider – the variation of the points on the scatter diagram around the regression line. As you will remember, in constructing the regression line we made the square of these deviations as small as possible. We minimised $\Sigma(Y - \hat{Y})^2$. Think about these variations for a minute. The value of Y is an established fact; sales reached a particular level and this we must accept. The value of $\hat{Y}$ we calculated from the regression equation and we have argued that the variation of $\hat{Y}$ from $\bar{Y}$ is fully explained by that equation. But there remains a part of the variation of Y (equal to $Y - \hat{Y}$) which is unaccounted for. It is a part of the total variation which is not explained by the regression equation. It is, then, an unexplained variation: this is shown in Fig. 6.7(c). Summarising then, we have:

$$\text{total variation squared} = \Sigma(Y - \bar{Y})^2$$
$$\text{explained variation squared} = \Sigma(\hat{Y} - \bar{Y})^2$$
$$\text{unexplained variation squared} = \Sigma(Y - \hat{Y})^2$$

Suppose we had spent £85,000 on advertising and as a result actual sales had been £6,000,000. From our regression equation we would have estimated that sales (Y) would be £4,594,900. The total variation is the actual value of sales (Y) less the average value of sales (£5,200,000) which is equal to £800,000. Of this total variation, an amount equal to $\hat{Y} - \bar{Y}$ is explained by the regression equation equal to £4,594,900 − £5,200,000, i.e. £ − 605,100. We would have expected sales to be £605,100 below the average, whereas in fact they were £800,000 above the average. There is then a variation of £1,405,100 which we have not explained.

What we have done in fact is to divide total variation into two parts – that part which we can explain, and that part which remains unexplained.

total variation − explained variation + unexplained variation

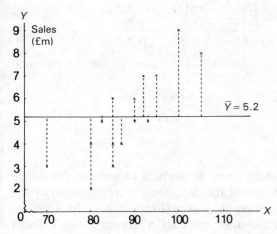

Figure 6.7a Total variation $\Sigma(Y - \bar{Y})^2$

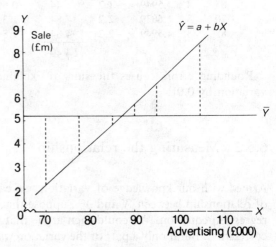

Figure 6.7b Explained variations $\Sigma(\hat{Y} - \bar{Y})^2$

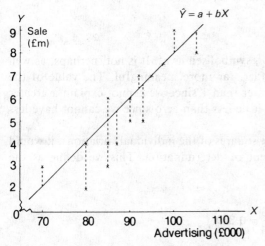

Figure 6.7c Unexplained variations $\Sigma(Y - \hat{Y})^2$

Regression and correlation 87

Let us now calculate these three variations from the data for Brown's Department Store.

				Total		Unexplained		Explained
X	Y	$\hat{Y}$	$Y - \bar{Y}$	$(Y - \bar{Y})^2$	$Y - \hat{Y}$	$(Y - \hat{Y})^2$	$\hat{Y} - \bar{Y}$	$(\hat{Y} - \bar{Y})^2$
100	9	7.5019	3.8	14.44	1.4981	2.2443	2.3019	5.299
105	8	8.4709	2.8	7.84	−0.4709	0.2217	3.2709	10.699
90	5	5.5639	−0.2	0.04	−0.5639	0.3180	0.3639	0.132
80	2	3.6259	−3.2	10.24	−1.6259	2.6436	−1.5741	2.478
80	4	3.6259	−1.2	1.44	0.3741	0.1400	−1.5741	2.478
85	6	4.5949	0.8	0.64	1.4051	1.9743	−0.6051	0.366
87	4	4.9825	−1.2	1.44	−0.9825	0.9653	−0.2175	0.047
92	7	5.9515	1.8	3.24	1.0485	1.0994	0.7515	0.565
90	6	5.5639	0.8	0.64	0.4361	0.1902	0.3639	0.132
95	7	6.5329	1.8	3.24	0.4671	0.2182	1.3329	1.777
93	5	6.1453	−0.2	0.04	−1.1453	1.3117	0.9453	0.894
85	5	4.5949	−0.2	0.04	−0.4051	0.1641	−0.6051	0.366
85	4	4.5949	−1.2	1.44	−0.5949	0.3540	−0.6051	0.366
70	3	1.6879	−2.2	4.84	1.3121	1.7216	−3.5121	12.335
85	3	4.5949	−2.2	4.84	−1.5949	2.5437	−0.6051	0.366
$\bar{Y} = 5.2$				54.40		16.1101		38.3

Rounding errors causes the sum of explained plus unexplained variation to differ from total variation by 0.01.

6.5.2 Measuring the relationship

Armed with our knowledge of variation, we can now go ahead to devise a measure of the degree of relationship between X and Y. Suppose that the whole of the variation Y were explained by the regression equation. We could then argue that the relationship was perfect − that we have perfect correlation. But if only a part of the variation was explained, the correlation between X and Y would be less than perfect. As a first approach then we could use as a measure of the relationship between the two variables that proportion of the variation in the values of the dependent variable which is explained. Thus, we have calculated the total variation squared to be 54.4 and the explained variations squared to be 38.3. As a first measure we have:

$$\frac{\Sigma(\hat{Y} - \bar{Y})^2}{\Sigma(Y - \bar{Y})^2} = 0.704$$

We call this measure the *coefficient of determination*, symbolised as r^2. It is not, perhaps, as well known as the *coefficient of correlation* but it is, in fact, far more meaningful. The value of the coefficient of determination cannot, of course, be greater than 1 since we cannot explain a greater proportion of total variation than the whole; nor can it be less than zero since we cannot have less than no variation explained.

Since we measured variation by taking the sum of the squares of the individual variations, it would seem logical to take the square root of the coefficient of determination. This we define as the coefficient of correlation.

$$r = \sqrt{\frac{\text{explained variation}}{\text{total variation}}} = \sqrt{\frac{\Sigma(\hat{Y} - \bar{Y})^2}{\Sigma(Y - \bar{Y})^2}} = \sqrt{\frac{38.3}{54.4}} = 0.839$$

Thus the coefficient is an abstract measure of the relationship between variables based on a scale ranging between $+1$ and -1.† Whether r is positive or negative depends on the nature of the relationship between X and Y. If the sign of b in the regression equation is $+$, as X increases so does Y, and we have positive correlation. On the other hand, if the sign of b is negative, you will find that as X increases the value of Y decreases and we have negative correlation.

Now there is nothing wrong in calculating the coefficient of correlation in this way, but it is, you must agree, a most cumbersome way of doing it. Not only have you to calculate the regression equation first, you have then to use that equation to obtain the values of $\hat{Y}$, obtain $\Sigma(\hat{Y} - \bar{Y})^2$ and then express this as a proportion of total variation squared. Like all such methods, it has its uses, and explains precisely what we are doing. But what we need is a method of calculating the coefficient of correlation directly without calculating the regression equation first.

6.5.3   Pearson's product moment correlation coefficient

We can calculate the value of the coefficient of correlation far more simply if we use the formula:

$$r = \frac{\text{covariance}(XY)}{\sqrt{\sigma_x^2 \times \sigma_y^2}} = \frac{s_{xy}^2}{\sqrt{s_x^2 \times s_y^2}}$$

As you will remember, the covariance (XY) is:

$$\frac{\Sigma XY}{n} - \frac{\Sigma X \cdot \Sigma Y}{n \cdot n}$$

The variance of X is:

$$\frac{\Sigma X^2}{n} - \left(\frac{\Sigma X}{n}\right)^2$$

The variance of Y is:

$$\frac{\Sigma Y^2}{n} - \left(\frac{\Sigma Y}{n}\right)^2$$

You will notice the similarity between this formula and the one used earlier to calculate the value of b in the regression equation. In fact, the numerator is the same and one term in the denominator is the same.

Let us now confirm that, in fact, this formula will yield the same coefficient of correlation as our previous calculations.

$$s_{xy}^2 = \frac{\Sigma XY}{n} - \frac{\Sigma XY \cdot \Sigma Y}{n \cdot n} = \frac{7{,}072}{15} - \frac{1{,}322 \times 78}{15 \times 15} = 13.17333$$

$$s_x^2 = \frac{\Sigma X^2}{n} - \left(\frac{\Sigma X}{n}\right)^2 = \frac{117{,}532}{15} - \left(\frac{1{,}322}{15}\right)^2 = 67.98222$$

$$s_y^2 = \frac{\Sigma Y^2}{n} - \left(\frac{\Sigma Y}{n}\right)^2 = \frac{460}{15} - \left(\frac{78}{15}\right)^2 = 3.62667$$

$$r = \frac{13.17333}{\sqrt{67.98222 \times 3.62667}} = 0.839$$

† Remember that $\sqrt{1} = \pm 1$.

6.5.4   The interpretation of r

One disturbing feature of many students' work is that although they can calculate the value of the coefficient of correlation in a perfectly efficient manner they completely ignore any effort to interpret what their results mean. What is the point of calculating a statistic if you cannot interpret it? We have calculated that r is $+0.839$ for the data we were working with. Thus it is clear that a strong positive correlation exists. But what does $+0.839$ mean? If we had another bivariate distribution for which $r = +0.42$, clearly our first coefficient implies a stronger correlation than the second. But is it twice as strong? In fact, it is far more meaningful to consider the coefficient of determination. Do you remember that this measures the proportion of total variation that can be explained by the regression equation? When $r = 0.839$, $r^2 = 0.704$ and we can conclude that 70.4 per cent of the variations in Y can be explained by the regression equation, leaving less than 30 per cent to be explained by other factors. But if we consider a coefficient of correlation of 0.42, we have a coefficient of determination of $0.42^2 = 0.1764$, which implies that only 17.64 per cent of the variations in Y are explained by the regression equation. Thus a coefficient of correlation of 0.839 is four times as strong as one of 0.42.

We would like, however, to issue a few words of warning about the interpretation of even a high coefficient. Even though analysis indicates that correlation exists, we are not justified in assuming that there is therefore a cause and effect relationship. We must never fall into the trap of assuming that cause and effect exists when it is nonsense to do so. Sometimes a high correlation is obviously nonsensical. There is, as noted earlier, a high positive correlation between the annual issue of television licences and the annual admissions into mental institutions. It would be ludicrous to suggest that cause and effect exists here. The only logical conclusion that one can draw is that, quite by chance, both statistics were increasing at the same rate. In this case it is quite obvious that cause and effect is not proved by a high correlation coefficient – but sometimes things are not quite so clear cut. The high correlation between infant mortality and the extent of overcrowding that was found in Bethnal Green between the First and Second World Wars does seem to suggest that overcrowding causes high infant mortality. In fact, however, both are probably a result of low income levels.

If, then, we can never use correlation analysis to prove a cause and effect relationship, you may wonder if it is worth study. Well, we think it is. Correlation can give added weight to a relationship that theory suggests exists. We could use it, for example, to verify the economic theory that consumption depends on income. Again, correlation can sometimes suggest that causal relationships exist in areas that were not previously suspected – a technique much used in medical research. Here, correlation can often point to promising avenues of investigation.

Finally, we would like to point out that a low correlation coefficient does not necessarily mean a low degree of association. After all, the coefficient we have calculated measures the strength of a linear relationship. The relationship may be very high, but curvilinear, and our coefficient would not indicate this.

6.5.5   Rank correlation

Positive correlation implies that large values of x are associated with large values of y and small values of x with small values of y. The converse is true, of course, of negative correlation. A useful method of assessing whether the data show any correlation would be to *rank* them, i.e. list the values of the variables as 1st, 2nd, 3rd and so on, the lowest rank being assigned to the lowest value of the variable.

Example

The following data refers to gross investment as a percentage of national income and the percentage growth in national income in nine selected countries. We are trying to assess if the level of investment is associated with the growth in national income. In order to make an initial judgement, the data has been ranked and the difference in the ranks of the two series listed in the column headed D.

	Gross investment (%)	National income growth (%)	Ranked			
	x	y	x	y	D	D^2
Belgium	19.2	3.4	4	2	2	4
Denmark	18.8	3.6	3	4	-1	1
France	19.5	4.9	5	$6\frac{1}{2}$	$-1\frac{1}{2}$	2.25
Germany	25.9	7.1	7	9	-2	4
Italy	21.8	5.6	6	8	-2	4
Netherlands	26.0	4.9	8	$6\frac{1}{2}$	$1\frac{1}{2}$	2.25
Norway	28.1	3.8	9	5	4	16
United Kingdom	16.4	2.6	1	1	0	0
United States	18.4	3.5	2	3	-1	1
						34.5

You will notice that in ranking the growth of national income two countries have a growth rate of 4.9 per cent. We cannot rank them 6th and 7th, or even equal 7th. Since the two countries are equal and between them occupy both the 6th and 7th ranks, we rank them equal at $6\frac{1}{2} = (6 + 7)/2$. Now looking at the ranks it does seem that there is some positive correlation here. While only the United Kingdom has the same rank in both series, only in the case of Norway is there any marked difference in the ranks. Generally, countries with a higher level of gross investment tend to have the higher growth rate.

We can, in fact, get a very good approximation to the correlation coefficient using ranks rather than the original data. Moreover, the calculations are easy as the formula simplifies to:

$$P = 1 - \frac{6 \sum D^2}{n(n^2 - 1)}$$

where D is the difference in ranks and n the number of observations. This is known as *Spearman's rank correlation coefficient* and in this case

$$P = 1 - \frac{6 \times 34.5}{9(81 - 1)} = 1 - \frac{207}{720} = 0.7125$$

Thus there is a positive correlation, but since $P^2 = 0.5076$, only about 50 per cent of the variation of y is explained. Correlation is not strong. Our knowledge of economics leads us to think that possibly we are examining the wrong variables. It may be net investment rather than gross that affects growth. This too could account for the difference in ranks in the case of Norway. Possibly much of the high level of investment here is replacement of worn-out equipment, or possibly it is investment which does not lead directly to growth.

One great advantage of rank correlation is that it can be used to rank *attributes* which cannot be given a numerical value. Thus we could assess the consistency with which different panels of judges assess the contestants in beauty contests such as Miss World. In the same way, we could assess whether different methods of selecting applicants for employment are likely to lead to the same results. Do we, for example, end up with the same ranking from a personal interview as we would from a written test or a psychological test? Thus we can use rank correlation in many cases where Pearson's correlation coefficient is not applicable.

Finally, notice that in the table $\Sigma D = 0$. This must be so (do you see why?) and provides a useful check on our initial calculations.

Review questions

6.1 From the following information draw the scatter diagram, and by the method of least squares draw the regression line of best fit.

volume of sales (thousands of units)	5	6	7	8	9	10
total expenses (£000)	74	77	82	86	92	95

What will be the expected total expenses when volume of sales is 7,500 units?
If the selling price per unit is £11, at what volume of sales will the total income from sales equal the total expenses?
Draw the total revenue line on your scatter diagram and verify your answer.

6.2 The projected sales of a certain company over the next 10 years are:

Year	1	2	3	4	5	6	7	8	9	10
Sales (£000)	3,221	2,944	2,697	2,478	2,284	2,087	1,919	1,766	1,608	1,486

Fit the curve $y = ab^x$ to this data. What can you learn about the value of b?

6.3 Sales of groggets for each of the past five years have been as follows:

year ending 31 May 19-2	12,500
year ending 31 May 19-3	15,480
year ending 31 May 19-4	18,640
year ending 31 May 19-5	21,480
year ending 31 May 19-6	24,400

The pattern of sales is represented by the quarterly indices 70, 90, 160, 80.
You are required to use the least squares method to determine the trend line of the company's sales, project the trend line into the following year and calculate the expected sales for each quarter of that year.

6.4 It is often suggested that low unemployment and a low rate of wage inflation cannot co-exist. Examine the evidence below and discuss whether it supports the above contention, producing the relevant statistical measures.

Year	Unemployment (%)	Change in wages (%)
1	1.6	5.0
2	2.2	3.2
3	2.3	2.7
4	1.7	2.1
5	1.6	4.1
6	2.1	2.7
7	2.6	2.9
8	1.7	4.6
9	1.5	3.5
10	1.6	4.4

Does more recent experience agree with your conclusions above?

11	2.5	4.0
12	2.5	7.7
13	2.5	5.7
14	2.7	9.5

6.5 The table below shows the rate of growth of national income and the level of gross investment as a percentage of national income in a number of countries.

Country	Gross investment (%)	National income growth rate (%)
Belgium	19.2	3.4
Denmark	18.8	3.6
France	19.5	4.9
Germany	25.9	7.1
Italy	21.8	5.6
Netherlands	26.0	4.9
Norway	28.1	3.8
United Kingdom	16.4	2.6
United States	18.4	3.5

(a) Rank each country by the level of investment and rate of growth.

(b) Calculate the rank correlation coefficient.

(c) If you were the Norwegian minister of economics, what comment would you have to make on this data?

Part Three   Sampling, estimation and significance

7 Sample design

7.1 Introduction

You will probably be aware that most statistical information is obtained, not by examining the whole population, but by obtaining a sample and arguing that the characteristics of the sample are the same as those of the population as a whole. The sample result, obviously, will differ from that we obtain by taking a complete census, and the next chapter will be concerned with the reliability of the sample results. Our objective here is not to discuss the results of a sample survey, but to explain to you how to select a sample which is truly representative of the population. You would not, for example, expect to obtain a reliable estimate of the heights of men in the United Kingdom if you chose your sample only from men in the Guards regiments; you could not possibly estimate the number of miles people travel by taxi each week if your sample consists only of taxi-drivers.

Such statements as this are, we think, obvious, but they only tell us how not to select a sample; they give no information about how we should select one. So let us proceed step by step. The basic principle underlying any selection of a sample is that it should be a *random* selection – that is, that every single unit within the population should have an equal chance (or the same probability) of being chosen as a member of the sample. This is, of course, very difficult to achieve. Consider the simple case of drawing a raffle ticket to select the winners of five prizes. Why do you think that after each ticket is drawn the drum containing them is turned? It is, as you will know, to mix up the tickets once again. But would it made any difference if the drum were not turned? In our opinion, it would make a great deal of difference. The person drawing the ticket can easily introduce *bias* into the selection. Consider carefully how you yourself would select the winning tickets. Out observation shows that few people will choose tickets from the top of the drum. They tend rather to plunge into the pile of tickets and select one from near the bottom. Can you see that this sort of behaviour means that tickets near the top of the pile do not have an equal chance of selection with those nearer the bottom? The drum is turned after each ticket is selected in order to offset this type of bias. We cannot change human behaviour, but we can change the position of the tickets so as to give each one an equal chance of selection.

To return to sampling proper, in selecting a sample to determine the average number of miles travelled by taxi each week, the person who never takes a taxi must have just as much chance of being selected as those who use taxis regularly – but no greater chance. In order to ensure this, statisticians have devised various methods of selection which, as far as possible, eliminate bias. If we are able to list the population in some sort of order and number them, we could determine the members of our sample by drawing numbers at random, drawing sufficient numbers to include, say, 5 per cent of the total population. Better still, we could use a table of *random numbers* to determine who should be included in our sample. This table is so constructed that if you select any point at which to start and move through it consistently in one direction, horizontally, vertically or diagonally, the digits you read off are randomly chosen and every number has the same probability of being selected. Suppose that the total population consisted of 1,000 units. These could be numbered from 0 to 999 and you could select those to be included in the sample by reading off digits from the table of random numbers in groups of three. If, for example, the first three digits were 294, we would select the unit numbered 294 as part of our sample. Similarly, if the first three digits were 004, we would select the unit numbered 4.

If, on the other hand, we were examining industrial output for faulty units, we could use a form of pseudo-random sampling, by examining and checking the unit coming off the production line at predetermined time intervals. Alternatively, we could examine, say, five units, chosen at random from every carton of output. One way or another, however, we must ensure that our selection of the sample is random.

7.2 The sampling frame

Before we can draw a truly random sample, we must be able to define the total population. The Electoral Register contains the names of all those entitled to vote in the United Kingdom; enrolment forms will list the names of all students in a certain college; business files will contain all the orders placed in a given financial year. Such lists, files or card indexes form what is known as a *sampling frame*. They will give basic details about every member of the population you are concerned with. Now, before drawing a sample, you must carefully examine the sampling frame to ensure that it is adequate for your purpose. If it is deficient in that a number of units are not included, those units have no chance of being selected and your sample cannot be truly random. Let us illustrate such deficiencies by looking at a very real problem.

It is the habit in Britain for public opinion polls to try to predict the results of government elections by interviewing a relatively small number of voters. The sampling frame for such a survey is the Electoral Roll, a list of all those eligible to vote who lived in the area last November. The list is produced in March each year, so when it is published it is already four months out of date. By the following February then, just before a new list is published, it is sixteen months out of date. Since about 0.5 per cent of the population move into and out of the district each month (according to a Social Survey report) this would mean that about 8 per cent of those who are on the Roll no longer live in the area, while about 8 per cent of those living in the area are not on the Roll. Add to this the fact that 4 per cent of those who should be on the register are omitted because they do not return the necessary forms, and we find that the Electoral Roll is only about 88 per cent accurate – not a good situation for a sampling frame. It is for this reason that investigators interview only those who actually go to the polling booth, or say that they intend to do so.

7.3 Systematic sampling

In practice, true random sampling is not possible unless there is a good sampling frame and the population is fairly small. So, for most practical work, investigators resort to methods of selection which are *quasi-random*, or not truly random. One of the most popular is to choose as a member of the sample every *n*th item on a list, the first sample unit being selected by some random method. If, for example, we wanted a sample consisting of $2\frac{1}{2}$ per cent of the population, we would first select at random the first unit, say the 29th invoice in a file, and then every 40th invoice after that – the

69th, 109th and so on. Can you see why this type of sampling is not truly random? While the first item is chosen by random methods, every other item is then preselected. Nevertheless, this method of selection (called systematic sampling) approximates to random sampling sufficiently well to justify its widespread use. Once again we need a reasonably good sampling frame but, in our opinion, having accepted that the sample is not truly random, it need not be 100 per cent accurate. However, if we are to use systematic sampling, we must ensure that there are no 'cyclical' patterns in the sampling frame that match our choice of items. If, for example, we are taking a survey on a housing estate to assess the opinion of the residents about the 'open plan' layout, and every tenth house is a corner house, then a 10 per cent systematic sample either always, or never, includes a corner house. Such a pattern as this could substantially affect the results we get.

Now, systematic sampling is useful so long as the population is homogeneous. It is suitable, for example, if accountants are selecting a sample of invoices to check for errors, or if the police are taking a sample of motorists passing a certain point to check that they have a valid driving licence. But you will appreciate that in much statistical work the population is heterogeneous. This is especially true if we are investigating opinions, because in many cases a person's opinions are formed by, or result from, the social class to which he or she belongs. If we are asking the simple question, 'Do you think that people earning over £25,000 a year pay too much in taxes? the answer we get will often depend on the income of the person questioned. Inevitably the rich will answer 'yes', while the poor are more likely to say 'no'.

7.4 Stratified sampling

When the response we get to our question is likely to depend in this way on the social group to which the person being questioned belongs, we will get far better results if we adopt *stratified sampling*. If we are investigating the social evils of traffic on the roads, it is quite possible that systematic sampling would result in a sample which contained only those who owned cars. The opinions of those who do not own cars would not be represented, and to that extent the sample result will be biased. Stratified sampling has been designed to ensure that all important views are represented in the sample. In this type of sample, each social group is represented in proportion to the size of that group in the population as a whole. Let us take a simple example to show how stratified samples are constructed. Suppose we are investigating opinions on education of those who are still at school or college. We may decide that these opinions will depend on whether the student is at university, state school or college. We know that 20 per cent of students are at university, 10 per cent at private school and 70 per cent at state school. If we are selecting a sample of 2,000 students, our first step is to say that 20 per cent of the sample, or 400 students, must be university students, 200 (10 per cent) from private schools, and 1,400 (70 per cent) from state schools. We may now additionally decide that men and women have markedly different opinions, so we will now divide our three groups (or strata) into two subgroups. If, for example, we know that 15 per cent of university students are women, we would include 60 women among our sample of 400 university students. Each of these subgroups can be further subdivided, perhaps in relation to age or to parents' social class. It does not matter how far groups are subdivided, provided that we are stratifying according to a characteristic which is relevant to the survey. If, for example, we are investigating television viewing habits, there is little point in stratifying the sample according to political affiliations; it is probably completely irrelevant. On the other hand, our viewing habits might well be affected by the size of our family or whether we live in a rural or urban area, and we would stratify according to these criteria.

Once we have decided on the number of people in each stratum of our sample, the persons to be included must, of course, now be chosen by some random method.

Now, one peculiarity can arise with this type of sampling. If our strata are many, it may be that an important subgroup of the population is entitled to be represented by only one or two members. It may be that we feel that we cannot obtain an adequate representation of the opinions of this group if we question one person and, if so, we should include more representatives, say four or five. While this may seem reasonable, it means, of course, that members of this group have a far higher probability of being chosen as part of the sample than have people in other strata. This type of sampling is known as sampling with a *variable sampling fraction*, and in assessing our final results we must make allowance for the difference in the representation of such strata. In many surveys, when samples are drawn from a markedly heterogeneous population, the sampling fraction is variable rather than uniform. The principle finds its most important application in the evaluation of stock value by sampling methods. We can afford to take a very small sample of items costing only a few pence since, if our sample is inaccurate, it will make little difference to the total value of stock. But if the items of stock cost a great deal, even a small error may significantly affect our total stock valuation. Thus, in determining total stock value from a sample valuation of stock, accountants almost invariably ensure that high-value items have a greater probability of being selected than do low-value items.

Once again, with stratified sampling, we need a sampling frame which will give us great deal of information about the social structure. Fortunately, nowadays governments undertake so many social surveys and publish so many statistics that there is little difficulty in obtaining the necessary information. Hence stratified sampling is becoming increasingly popular whenever the objective is to assess people's attitudes or opinions.

7.5 Multi-stage sampling

One of the problems of simple random sampling is that if the sample selected is widely scattered over the country the interviewer may spend more time in travelling than he does in actually interviewing. Often, then, the cost of taking such a sample can be prohibitive. This is especially true in countries in which urbanisation has not yet developed. You can imagine the problems involved if you were to take a sample of people scattered all over central Canada, Malaysia or West Africa. Interviewers might have to spend weeks in travelling.

It is to overcome this problem that multi-stage sampling has been developed. Rather than spending time in travelling to interview 2,000 people scattered all over Malaysia, is it not more convenient to interview say 100 people in each of 20 selected areas? Provided that we can ensure that the sample ultimately selected is representative, a great deal of time and expense can be saved.

In many ways the selection of a multi-stage sample is similar to the selection of a stratified sample, but we select primary groups and subgroups geographically rather than on the basis of social characteristics. Typically, the first stage is to break down the area under survey into a number of standard regions. These may be areas such as the English county, the Canadian province or any other easily defined administrative area. The sample is then divided among these regions according to their population. The second stage is to select at random a small number of districts, say towns and villages, within the primary region. Once again it is necessary to allocate to each of these districts a number of interviews proportional to its population. Almost certainly at this stage, an element of stratification must appear if the sample is to be representative. Let us suppose that we are selecting six towns at random within each region: If a third of the population live in cities of over 200,000 people, a further third in towns with a population of from 20,000 to 200,000, and the remainder in small towns and villages of under 20,000 inhabitants, it would be desirable to select two areas of each type.

Finally, having selected our towns and villages within each region, the sample in each town or village is chosen by some random method. Again, it may be felt to be desirable to choose a stratified sample within each town. How far this stratification is carried depends, of course, on the purpose of the survey and the homogeneity of the population. If we are examining housing conditions, and the type of housing ranges from one-roomed flats to 25-roomed mansions, some degree of stratification will be necessary. But if all the housing in the area is of an essentially similar type, as it may be in a small village, stratification is probably not necessary.

Every type of sampling discussed so far depends on the existence of a sampling frame. If no such frame exists, random sampling is not possible, and to undertake a census to derive a sampling frame is a very expensive process indeed. Unfortunately, in most underdeveloped countries, satisfactory sampling frames do not yet exist. Sometimes too, the very cost of conducting a survey may make it necessary to use a different type of sampling. To overcome these problems statisticians have designed two further sampling methods – cluster sampling and quota sampling.

7.6 Cluster sampling

Cluster sampling was devised in the United States to try to overcome the problems of cost and the lack of a satisfactory sampling frame. Instead of selecting a random sample scattered over a wide area, it is a few geographical areas that are selected at random and every single household in each area is interviewed. Obviously the areas chosen must be relatively small. It is not possible to interview every household in cities like London and Singapore. The typical area selected might be three or four streets in a town, or perhaps an apartment block. The great advantage of this type of sampling is the saving in time and cost. Many interviews can take place within a short space of time with a minimum of travelling. Moreover it does not require any knowledge of the population before the survey is undertaken; that is, no sampling frame is necessary.

On the other hand, there is a basic problem with this type of sample. While the population is heterogeneous, those who live in a small area or the same block of flats will tend to be homogeneous. They will tend to have the same opinions, the same characteristics, the same life-style. Naturally one cannot be certain of this, but it seems to be highly probable, and since the statistician is selecting only a few such areas there is a great danger that the sample will be biased. One way of trying to offset this tendency to bias is to increase the number of clusters in the hope that you will then include in your sample every important strata, but every increase in the number of clusters raises the cost of the survey. There comes a time when you will have to weigh the risk of selecting a biased sample against the cost of avoiding the danger of bias. It is, however, a useful type of sample when you have no sampling frame, or when the cost of taking a random sample is too high.

7.7 Quota sampling

There can be few people who have not been stopped in the street by an interviewer holding a questionnaire, or who at least know someone who has been so stopped. And the immediate reaction seems to be, 'Why did they choose me?'

The essence of this type of sample is that it is not preselected, but is chosen by the interviewer on the spot. He, or more usually she, is given a certain number (a quota) of questionnaires which it is her job to have completed during the course of the week or month. She has a free choice of whom to ask to answer the questions. This choice, naturally, is subject to some general restrictions. One can imagine a young male interviewer stopping only girls of about his own age whom he thought to be pretty. Free choice does not extend as far as this. The interviewer is told that the completed quota of questionnaires must include a certain number of males and of females; a certain number from specified age groups, such as under 18, 18 to 40, 40 to 65. Controls such as this are easy to implement, but often the quota has to be further subdivided by social class or occupation. This can make quota sampling very difficult as there is no real definition of what constitutes, say, the upper middle class, and certainly you can seldom tell a person's social class by looking at him/her. Such a requirement necessitates the interviewer being given detailed definitions and descriptions of what the survey body means by each term it uses. Subject, however, to controls such as this, the sample is chosen by the interviewer from people who pass in the street.

Now, obviously, such a technique of sampling can be open to a great deal of abuse and bias. When it was first introduced, many cases were reported of interviewers sitting at home filling in their own questionnaires without ever interviewing anyone. Of course, this was soon remedied by the questionnaire asking for names and addresses, and the survey body making a spot check to ensure that such people had actually been interviewed. Deliberate fraud of this nature is very rare now. Much more important is the bias to which the quota survey may be subject. If the interviewer does not start work until, say, ten o'clock, the majority of people he/she can stop must of necessity be the housewife or the unemployed; if he/she starts at, say nine o'clock, the vast majority will be clerical or managerial workers; to start even earlier means that he/she will probably meet largely manual workers. Thus, the interviewer must be prepared to spread work throughout the day. It is little use trying to have all the questionnaires completed before the morning coffee-break. A further source of bias is that people you meet in the street are either going somewhere or doing something; they resent being interrupted and their answers may be hurried and slapdash, given without serious thought. These problems necessitate further controls and advice being given to the interviewer. He/she may be told at what times to conduct the interviews or even where to conduct them. We would not wish to imply that interviewers are rascals or rogues – far from it – but we must point out that the success of such a survey depends very much on their following their instructions to the letter.

Since the quota sample has these weaknesses, why is it becoming increasingly common? Well, for one thing, it is cheap. It eliminates repeated calls to interview a person who may not be at home on the first two or three occasions. In fact, it is estimated that each interview in a quota sample costs only about half as much as each interview in a random sample. Then, too, we have in every survey a number of people who do not respond. In a quota sample, it does not matter. They are ignored and we pass straight on to the next person with no loss of time and no cost. In other types of sample, a substitute has to be found for each person who does not respond. It might, in fact, be argued that this itself introduces bias into the sample. It may be that the man who is prepared to fill in questionnaires is some kind of extrovert, and we may be getting the views of only one type of individual.

Nevertheless, given the existence of the controls and checks we have mentioned, all the evidence points to the fact that in skilled hands quota sampling gives reasonably satisfactory results. Statisticians generally accept that it is not a substitute for random sampling statistically, but it is so quick and so cheap a method of carrying out surveys that it is not likely to be replaced for a very long time.

Finally, we must stress that the effectiveness of any sample survey, whatever the method of sampling used, depends to a very large extent on the questions which are asked. If they are ambiguous, so will be our answers; if the interviewer puts the questions with a bias, our results will be biased. No survey result can be better than the questionnaire on which it is based.

8 Sampling and confidence limits

8.1 Introduction

You will have realised by now that the art of sampling is central to the study of statistics, and that we sample to learn something about the population from which the sample was drawn. As the sample we draw is likely to form only a fraction of the population, it is evident that any conclusion we draw about the population is subject to error and, as we saw in the previous chapter, we can minimise the scale of such errors by using sound sampling methods. However, despite all the care we may take in designing the sample, the chance of making errors still exists, and we shall now turn our attention to attempts to quantify the scale of such errors. Before we start, it must be stated that our analysis of sampling error will be limited in two ways. First, we will confine our attention to large samples, as small samples create problems of their own which we shall examine later. This naturally begs the question as to what constitutes a large sample, and we shall state now (without any justification) that a *large sample is a sample with more than 30 items*. The second limitation that we will place on our analysis of sampling is that we will confine our attention to simple random samples.

It is a fundamental truth about statistical theory that if we have a prior knowledge of the population, we can make predictions about the behaviour of samples drawn from it. Now you may object that this is of little use: we want to draw conclusions about the population from the sample. However, before we can assess the validity of such conclusions, it is necessary to understand thoroughly the theory of sample behaviour.

8.2 The sampling distribution of sample means

Let us start by assuming that we have a population with a known mean μ and a standard deviation σ. From this population we will draw a sample of n items, and for each sample we will find the sample mean $\bar{x}$. Clearly, each time we draw a sample, we are not going to obtain the same value for the sample mean, so it would be sensible to arrange our sample means in a frequency distribution. We call this frequency distribution *a sampling distribution of sample means*, and we shall now attempt to make some prediction about this sampling distribution.

Suppose you were asked to guess the value of one of these sample means, what would your guess be? As we know that the population mean is μ, it would surely be sensible to guess that the sample mean would also be μ. For example, if we know that the average height of all male adults in a certain town is 175 cm, then it would be sensible to guess that a sample of 100 men from this town would also have an average height of 175 cm. Now of course, in practice, any *individual* sample mean cannot be expected to be exactly the same as μ: some will be more than μ and some less than μ. So perhaps it would be more sensible to generalise with a statement about all sample means like this:

The average value of $\bar{x}$ (the sample mean) is μ (the population mean)

We can expect, then, that the sample mean will be clustered round the population mean, and it would be sensible to measure the degree of spread with a measure of dispersion. The measure we will use is the standard deviation, but you should note that to distinguish the fact that we are considering a sampling distribution rather than a sample or a population, in such cases we call the standard deviation the *standard error* (s.e). Now the size of the standard error will depend on the size of the sample drawn, and will lie within limits we can define precisely. Let us consider the largest sample we can possibly draw: this would occur if we selected every item within the population. The sample mean would be exactly equal to the population mean, and we would obviously obtain the same value for the sample mean in every sample we selected. So the sample mean would not show any spread, and the standard error of the mean would be zero. Now suppose we selected the smallest sample possible, i.e. one item only. The sample means would be the same as the value of the individual items drawn, and so the sample means would show the same spread as the individual items in the population. In this case the standard error of the mean would be σ, the population standard deviation. We can conclude, then, that the standard error of the mean must be at least zero, and at most σ. Moreover, as the sample size is increased, then the standard error decreases. In fact, it can be shown that:

The standard error of the mean $= \sigma/\sqrt{n}$

where n is the sample size.

Now, if the population is normally distributed, you would expect the sample means to be normally distributed, but what is rather curious is that, even if the population is not normally distributed the sample means will still turn out to be normally distributed, *as long as the sample is reasonably large*. This strange feature of the sample mean is called the *central limit theorem*, and explains why we spend so much time considering the normal distribution.

Example 1

What proportion of samples of 100 items drawn from a population mean of 100 and standard deviation of 20 would have a mean greater than 105? We would predict that the average value of the sample mean would be 100, and the standard error of the sample mean would be $20/\sqrt{100} = 2$. Figure 8.1 illustrates the proportion we require.

The Z score for 105 is:

$$\frac{105 - 100}{2} = 2.5$$

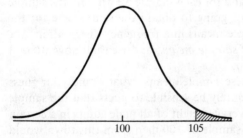

Figure 8.1

and consulting the normal distribution tables we can expect 0.621 per cent of sample means to exceed 105.

Example 2

A confectioner produces cakes which he packs in cartons of 100 for the catering trade. He prints on the carton 'average weight per cake not less than 95.5 grams'. If the cakes have a mean weight of 100 grams and a standard deviation of 20 grams, what proportion of batches contravene the Trade Description Act?

Every time the confectioner packs a carton, this is equivalent to selecting a sample of 100 cakes. We can consider, then, that the average weight of cakes in the carton forms a sampling distribution of sample means with a standard error $20/\sqrt{100} = 2$. As we wish to know the proportion of cartons with a mean weight of cakes less than 95.5 grams, the problem can be illustrated as in Fig. 8.2.

The Z score for 95.5 is:

$$\frac{95.5 - 100}{2} = -2.25$$

and consulting the tables we see that only 1.222 per cent of cartons would contravene the Act.

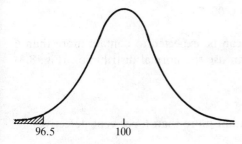

96.5   100

Figure 8.2

8.3 The sampling distribution of sample proportions

There has been an increasing tendency in recent years to give the results of sample surveys in percentage form. A manufacturer may claim that not more than 3 per cent of his output develops faults in the first year of use; public opinion surveys may report that a certain percentage of the population prefer filter tip cigarettes to plain cigarettes; or that 38.2 per cent of the population can be expected to vote Labour at the next general election. But most statisticians prefer to use proportions rather than percentages; they would report that the proportion that can be expected to vote Labour is 0.382.

Let us suppose that we are considering a population, some members of which possess a certain attribute. The population might be a large batch of manufactured goods, and the attribute might be that the good is defective. If we draw many samples from this population, calculate the proportion in each sample that has the attribute and record our results in a frequency distribution, then we have formed a sampling distribution of sample proportions. Now if we know the proportion in the population that has this attribute (call this proportion π), then we can make the following predictions about the sampling distribution:

The mean proportion in the sampling distribution will be π – the same as the proportion in the population.

If the samples all contain n items, then the standard error of the proportions is

$$\sqrt{\frac{\pi(1-\pi)}{n}}$$

Notice that, just like the standard error of the sample mean, the larger the sample size the smaller will be the standard error.

Example 3

Suppose we know from past experience that 5 per cent of a certain output is defective. We intend to draw samples of 800 items and find the proportion defective in each. In this case, we have $\pi = 0.05$, so

The mean proportion defective in the samples $= \pi = 0.05$

The standard error of proportions is $\sqrt{\dfrac{0.05(1-0.05)}{800}} = 0.0077$

If we wish to know the proportion of the samples that can be expected to contain more than 6 per cent defective items then, as the sample is large, we can use the normal distribution (Fig. 8.3)

$$Z = \frac{0.06 - 0.05}{0.0077} = 1.3$$

The table tells us that 9.68 per cent of samples would contain more than 6 per cent defectives.

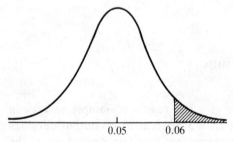

0.05   0.06

Figure 8.3

8.4 The sampling distribution of sums and differences

We shall now discuss sampling distributions formed by taking sums or differences. To understand just what this means consider modern industry: a considerable part of industrial activity is based, not on making something, but on assembling components that have been made in a number of different factories. This is especially true of the motor industry. Let us take a very simple case and imagine that we are joining end-to-end two straight metal bars produced in two workshops. The first shop aims to produce bars with lengths of 20 cm, but of course there will be variation in their actual lengths. We will assume that the lengths of bars are normally distributed with a mean of 20 cm and a standard deviation of 0.05 cm. Bars produced in the second workshop we will again assume to be normally distributed, this time with a mean length of 10 cm and a standard deviation of 0.04 cm.

When a workman picks up two bars for joining he is, in effect, sampling. He is drawing a sample of two items, one from the first workshop and one from the second. Almost certainly, the lengths of the assembled bars will be of importance to the firm. The length of any one assembled bar is obtained by taking the sum of their lengths, and just as the lengths of the individual bars varies, then so will the lengths of the assembled bars. We could, in fact, form a frequency distribution of the lengths of the assembled bars. As the bars have been obtained by *sampling*, and as the lengths have been obtained by *adding*, the frequency distribution of the lengths of assembled bars is called a *sampling distribution of sums*. What can we conclude about this distribution? Well, as the lengths of the individual bars are normally distributed it would seem reasonable to assume that the lengths of the assembled bars will also be normally distributed. It is true that

> *If the populations are normally distributed, so will be the distribution of sample sums formed from the populations.*

Suppose you were asked to guess the lengths of the assembled bars – it is probable that you would argue something like this: as the bars produced in the first workshop have a mean length of 20 cm and bars produced in the second workshop have a mean length of 10 cm, then the mean length of the assembled bar would be $20 + 10 = 30$ cm.

> *The mean of the sample sums is the sum of the means of the populations from which they were formed, i.e. mean of sample means $= \mu_1 + \mu_2 + \mu_3 + \cdots + \mu_\pi$.*

What can we conclude about the dispersion of the lengths of assembled bars? Suppose we take $\mu \pm 4\sigma$ as covering the effective limit of any normal distribution (consult the table and you will see that it covers all but 0.06 per cent). The longest bar produced in the first workshop would be $20 + 4 \times 0.05 = 20.2$ cm, and the shortest would be $20 - 4 \times 0.05 = 19.8$ cm. This gives range of lengths of $20.2 - 19.8 = 0.4$. Bars produced in the second workshop would have a maximum length of $10 + 4 \times 0.04 = 10.16$ cm, and a minimum length of $10 - 4 \times 0.04 = 9.84$ cm. This gives a range of $10.16 - 9.84 = 0.32$ cm. When a workman picks up a rod for joining he could pick the two longest bars and produce an assembled bar of $20.2 + 10.16 = 30.36$ cm. He could choose the two shortest bars and produce an assembled bar $19.8 + 9.84 = 29.64$ cm. So the range of assembled bars is $30.36 - 29.64 = 0.72$ cm. *Notice that this is the sum of the ranges of the two populations* $(0.4 + 0.32 = 0.72)$. However, we do not usually measure dispersion with the range – we prefer the standard deviation (as we are dealing with a sample, we should really call this measure the standard error). Unfortunately, what is true of the range is not true of the standard error. We cannot add the two populations' standard deviations and conclude that the standard error of the sample sum is $0.04 + 0.05 = 0.09$. But although we cannot add the standard deviations we can add the variances.

> *The variance of the sample sums is the sum of the variances of the populations from which they were formed.*

The additive property explains why statisticians often prefer to use the variance to the standard deviation. We can now conclude that

> *Standard error of the sample sum* $= \sqrt{\sigma_1^2 + \sigma_2^2 + \cdots + \sigma_\pi^2}$

The standard error of the lengths of the assembled bars would be

$$\sqrt{0.04^2 + 0.05^2} = 0.064$$

Example 4

A manufacturer produces two types of metal rods. Type A has a mean length of 3.5 cm and a standard deviation of 0.3 cm. Type B has a mean length of 4.5 cm and a standard deviation of 0.04 cm. A rod of each type is selected, joined together, and the joined rods are placed in a slot 8.1 cm long. What proportion of rods will not fit the slot?

Mean length of joined rods $= 3.5 + 4.5 = 8$ cm
Standard error of joined rods $= \sqrt{0.03^2 + 0.04^2} = 0.05$

The shaded area in Fig. 8.4 illustrates the proportion of assemblies that will not fit. To find this proportion, we must find the Z score for 8.1 cm.

$$Z = \frac{8.1 - 8}{0.05} = 2$$

Consulting the table we would expect that 2.275 per cent of assembled rods will not fit the slot.

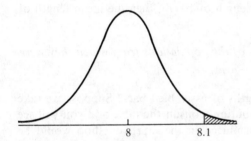

Figure 8.4

We shall now turn our attention to considering the sampling distribution of sample differences, and we shall do this by developing Example 3. Let us suppose that part of the assembly process of a certain machine involves placing a metal rod into a slot. The bars have lengths that are normally distributed with a mean of 8 cm and a standard deviation of 0.03 cm. You should check for yourself that all but 0.06 per cent of bars will have lengths within the limits 7.88 cm to 8.12 cm – a range of 0.24 cm. Slots have lengths that are normally distributed with a mean of 8.3 cm and a standard deviation of 0.04 cm, so slots will have lengths within the limits 8.14 cm. to 8.46 cm – a range of 0.32 cm. Now there will be a gap between the slot and the bar – it is measured by subtracting the bar length from the slot length. The gaps would then form the sampling distribution of sample differences. Now as you would imagine, the mean gap would be $8.3 - 8.0 = 0.3$ cm.

The mean of the sample difference is the difference between the means of the populations from which they were formed, i.e. mean of the sample differences $= \mu_1 - \mu_2$.

Turning now to the range of the gap, the largest gap will be obtained by taking the largest slot with the smallest bar, so the largest gap is $8.46 - 7.88 = 0.58$ cm. The smallest gap is obtained by taking the smallest slot with the largest bar, so the smallest gap is $8.14 - 8.12 = 0.02$ cm. So the range of gaps is $0.58 - 0.02 = 0.56$ cm. This is the sum of the ranges of the two populations

$(0.24 + 0.32 = 0.56)$. As the variances are additive we can conclude that

Standard error of the sample differences $= \sqrt{\sigma_1^2 + \sigma_2^2}$

Notice that even though the distribution is formed with a sample difference, we still add the variances.

Example 5

Full boxes of rice have weights that are normally distributed with a mean of 225 grams and a standard deviation of 1.5 grams. The empty boxes have weights that are normally distributed with a mean of 7 grams and a standard deviation 0.05 grams. Printed on the packet is the claim that the minimum net weight of contents is 214 grams. What proportion of packets fail to meet the claim?

Mean weight of contents $= 225 - 7 = 218$ grams
Standard error of weights of contents $= \sqrt{1.5^2 + 0.05^2} = \sqrt{2.2525}$

The shaded area in Fig. 8.5 represents the proportion we require. The Z score is

$$\frac{214 - 218}{\sqrt{2.2525}} = -2.665$$

The table tells us that a very low proportion – only 0.385 per cent – of packets are below the specified minimum weight.

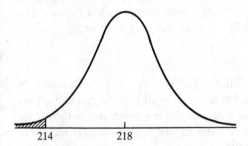

214   218

Figure 8.5

8.5   Confidence limits

Let us return to the example we considered in Example 1, i.e. the population with a mean of 100 and a standard deviation of 20. If we draw samples of 100 from this population, then the means will be normally distributed with a standard error $20/\sqrt{100} = 2$. Let us now ask between what limits we would expect the central 95 per cent of sample means to lie. Figure 8.6 illustrates the problem. We are excluding the 2.5 per cent at the extreme right of the distribution and the 2.5 per cent at the extreme left, and we require the Z scores for points a and b. Now b has a Z score that is exceeded only on 2.5 per cent of occasions, so consulting the table we find that we find that b must have a

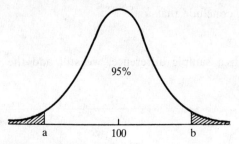

95%

a   100   b

Figure 8.6

Z score of 1.96. In other words:

$$\frac{b - 100}{2} = 1.96$$

so $b = 103.92$

As a and b are symmetrical about the mean, a must have a Z score of -1.96, so:

$$\frac{a - 100}{2} = -1.96$$

so $a = 96.08$

Thus, 95 per cent of sample means can be expected to fall within the range 96.08 to 103.92. *Such limits are known as confidence limits, and the one we have calculated is the 95 per cent confidence limits.* Now let us generalise. If we draw samples of n items from a population with a mean μ and a standard deviation σ, then 95 per cent confidence limits for sample means are given by

$$\mu \pm 1.96\sigma/\sqrt{n}$$

Beware of your interpretation of 95 per cent confidence limits. The concept does *not* mean that we are 95 per cent sure that a single sample mean lies within these limits – all that we can say is that either it will lie within these limits or it will not. *The 95 per cent confidence limits mean that if we drew many samples and find the mean for each then we can expect 95 per cent of sample means to be within the stated limits.* Now of course you may not be satisfied with 95 per cent confidence limits, and you may want to be more sure than this. We could, for example, use 99 per cent confidence limits for sample means, in which case the expression becomes:

$$\mu \pm 2.576\sigma/\sqrt{n}$$

So for our last example we can expect 99 per cent of sample means to lie within the range:

$$100 \pm 2.576 \times 20/\sqrt{100}$$

i.e from 94.848 to 105.152. So we notice that if we wish to increase the level of confidence we must widen the range for the sample mean. Although we have examined confidence limits for sample means, they could equally well apply to sample proportions. For example, the 95 per cent confidence limits for the sample proportion is

$$\pi \pm 1.96 \times \sqrt{\frac{\pi(1 - \pi)}{n}}$$

8.6 Summary

In this chapter, we have introduced concepts that are of fundamental importance in statistics, and it would be useful to summarise what we have learned. If we draw samples and derive some statistic (for example the sample mean) and record our results into a frequency distribution, then we have formed a sampling distribution. The standard error measures the variation in the sample statistics, and we met three standard errors:

standard error of the mean $= \sigma/\sqrt{n}$

standard error of the proportion $= \sqrt{\dfrac{\pi(1 - \pi)}{n}}$

standard error of the sum or difference $= \sqrt{\sigma_1^2 + \sigma_2^2}$

A confidence limit is a prediction, giving the central limits for sample results.

95 per cent confidence limit $= \pm 1.96 \times$ s.e.
99 per cent confidence limit $= \pm 2.576 \times$ s.e.

Review questions

8.1 Suppose that samples of 100 items were drawn from a population with a mean of 200 and a standard deviation of 30. What proportion of samples will have means

   (a) greater than 202,
   (b) less than 199,
   (c) between 198.5 and 203?

Set up 95 per cent and 99 per cent confidence for the sample mean.

8.2 A manufacturer produces metal rods which have a mean breaking strength of 60 kg and a standard deviation of 5 kg. He wishes to pack the rods in bundles and guarantee that the breaking strength of the rods exceeds 59 kg. How many rods should he pack in each bundle if he wishes no more than 5 per cent of bundles to violate the guarantee?

8.3 A machine cuts rods to a specified length of 2 cm and it is known that the machine operates with a standard deviation of 0.02 cm. If a sample of 100 rods has a mean length of 2.005 cm, what would you conclude?

8.4 It is known that some 8 per cent of components produced are defective. If components are packed in batches of 1,000, what proportion of batches contain more than 8.5 per cent defective items?

8.5 Set up 95 per cent and 99 per cent confidence limits for the proportion defective in samples of 500 drawn from a population that contains 7 per cent defective items.

8.6 The Ruritanian Ministry of Agriculture has set a standard 94 per cent germination rate for grain seed, and only 2 per cent of batches of seed can have germination rates less than the standard. If a certain crop of grain seed is known to have a 95 per cent germination rate, how many seeds should be in each batch in order to meet the minimum requirement?

8.7 There are three distinct stages in the servicing of a particular machine.

Stage	Mean time (min)	Standard deviation (min)
Cleaning	18	1.0
Greasing	12	0.5
Setting	10	0.5

Assuming that the times are independent and normally distributed, find the proportion of machines serviced in less than 37.5 minutes.

8.8 Now you will have to think about this one! Part of the assembly of a machine involves placing a metal rod into a slot. Both the rods and the slots have lengths that are normally distributed with the following parameters:

	Mean	Standard deviation
Rods	16.0 cm	0.03 cm
Slots	16.1 cm	0.04 cm

Find the distribution of the clearance between the bars and the slots. If a clearance of under 0.02 cm is satisfactory, find the proportion of assemblies that fail to meet this requirement.

9 Estimation

9.1 Introduction

In the previous chapter, we concerned ourselves with making predictions about sample behaviour, and we did this by introducing the idea of sampling distributions and confidence limits. In order to form sampling distributions, we required knowledge of the appropriate population parameters. If we were variate sampling, then we needed to know the population mean and variance; and if we were attribute sampling, then we needed to know the population proportion. We will now reverse the process: we will assume that the population parameters are unknown to us and that we will want to predict them from sample evidence. We will use the technique known as *statistical estimation*. It goes without saying that we require any estimate made to be a 'good' one – an estimate that a statistician would call *unbiased*. Before we start, then, we need to define precisely what is meant by an unbiased estimate. *An unbiased estimate is one that is just as likely to be too great as it is to be too small.* In other words, if we drew all possible samples of size n and made an estimate of the population parameter from each, then the average value of our estimate would be equal to the true value of the population parameter. Obviously this is a sensible definition – the more data that we have, then the more accurate would be our estimate.

9.2 Estimating a population mean

Let us suppose that we draw a sample of *n* items, and we use the sample to estimate the population mean. We can use the symbol ^ to mean 'an estimate of', so the estimated population mean is denoted by $\hat{\mu}$. The unbiased estimate of the population mean will be the sample mean, so:

$$\hat{\mu} = \bar{x}$$

This single estimate of μ is called a *point estimate*, and we are going to be extremely lucky if our estimate is bang on target. It would seem more reasonable to state a confidence interval, or *interval estimate* of the population mean.

To set up this confidence interval, we must decide on how sure we want to be that our interval estimate does include the true value of the mean and, as we saw in the previous chapter, it is usual to apply a 95 per cent level of confidence. To see how the method works, let us suppose that we know that a population has a mean of 100 and a standard deviation of 20, though we keep this information to ourselves. We now ask a statistician to estimate the mean of the population by sampling. If the statistician decides to draw 100 items for his/her samples, we can use confidence limits to predict that

95 per cent of the means of such samples would be in the range:

$$100 \pm 1.96 \times 20/\sqrt{100}$$

or (roughly) within the range 96 to 104.

The statistician now draws his/her samples, and sets up a 95 per cent interval estimate from each sample. Suppose the first sample had a mean of 99.5, then his/her interval estimate for the population mean would be:

$$\hat{\mu} = 99.5 \pm 1.96 \times 20/\sqrt{100}$$

or between 95.5 and 103.5, an interval estimate that does in fact bracket the true value of the mean. If his/her second sample had a mean of 102.5, then the interval estimate would be

$$\hat{\mu} = 102.5 \pm 1.96 \times 20/\sqrt{100}$$

or between 98.5 and 106.5, which again brackets the true value of the population mean. Figure 9.1 illustrates the situation, and a glance at it will show that provided the sample mean lies between 96 and 104, then the interval estimate will bracket the true value of the population mean. Now sampling theory tells us that we can expect 95 per cent of sample means to lie within these limits, so we can expect 95 per cent of such interval estimates to bracket the true value of the mean. This is what is meant, then, when we talk of a 95 per cent interval estimate of the population mean. If we draw many samples from a population, and set up interval estimates for the population mean, then 95 per cent of our interval estimates will include the true value of the population mean.

Now if you think carefully about what we have learnt so far, you will realise that there are two snags with this analysis. First, we have used the population standard deviation to set up an interval estimate for the population mean, but in practice we will not know the true value of the population standard deviation, and will have to use the sample standard deviation (s). Our interval estimate is, then:

$$\hat{\mu} = \bar{x} \pm 1.96s/\sqrt{n}$$

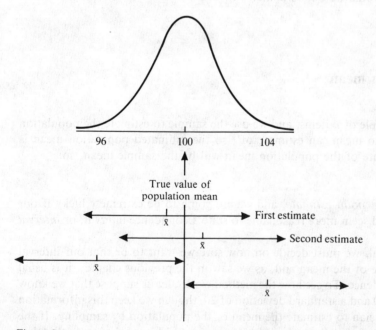

Figure 9.1

Now as s is only an estimate of σ (the population standard deviation), to retain 95 per cent confidence, we should really widen our estimate a little. However, you can rest assured that as long as our sample is large (greater than 30), the amount by which we would have to widen our estimate is insignificantly small.

Second, we have assumed that the statistician has drawn many samples to estimate the population mean, but in practice he/she will draw just one. Once the estimate has been made, then the statistician is either right or wrong – there is no way he/she can be right on 95 per cent of occasions. What, then, would an interval estimate mean in this case? The statistician is using a method that has a 95 per cent chance of success in the long run: he/she knows that if he/she uses this method every time he/she is asked to estimate a population mean, then 95 per cent of interval estimates will bracket the true value of the mean. He/she will be right on 95 per cent of occasions.

9.3 Estimating a population proportion

Quite often, we will find it useful to use a sample to estimate the proportion within a population with a certain characteristic (π). For example, we may wish to estimate the proportion in the population who intend to buy a certain product, or the proportion of defective items in a large batch. To do this, we can use a method essentially similar to estimating a population mean, i.e. we can set up a 95 per cent interval estimate. Now the point estimate of the population proportion is the sample proportion (which we shall call p), so

$$\hat{\pi} = p$$

To calculate an interval estimate, we will need the standard error of the proportion. Using the sample proportion, this is

$$\sqrt{\frac{p(1-p)}{n}}$$

So our 95 per cent interval estimate for the population proportion is:

$$\hat{\pi} = p \pm 1.96 \sqrt{\frac{p(1-p)}{n}}$$

Example 1

In order to estimate its share of the market, a detergent manufacturer randomly selects 500 women and asks them whether they use the manufacturer's products. Two hundred and sixty report that they do so. Estimate the manufacturer's market share.

$$\pi = p = 260/500 = 0.52$$

The 95 per cent interval estimate for the market share is:

$$0.52 \pm 1.96 \sqrt{\frac{0.52(1-0.52)}{500}}$$

$$= 0.52 \pm 0.0437$$

or, using percentages:

52% ± 4.37%

9.4 Determining the sample size

We can see that we can give a 95 per cent interval estimate for a population parameter by calculating

sample statistic ± 1.96 × standard error

the quantity ± 1.96 × standard error being called the *sampling error*. Now clearly, the smaller the size of the sampling error, the more precise will be the interval estimate. We shall now ask what determines the size of the sampling error, and see what we can do to reduce it. If we multiply the standard error by a constant smaller than 1.96 then we will reduce the sampling error − but we will no longer have the interval estimate we require as our level of confidence would be less than 95 per cent. If, then, we are to reduce the sampling error we must turn our attention to the standard error.

Let us turn our attention first to estimating the population mean. In this case, the standard error is $s/\sqrt{n}$, and if we reduce the size of s or increase the size of n, then the standard error (and so the sampling error) would be reduced. Now although we cannot reduce the size of s (after all, the sample standard deviation is quite outside of our control), we can increase the size of n. This will improve the precision of our estimate. Think about this for a minute and you will realise that it makes sense, because after all we should have more confidence in large sample results than in small sample results.

Suppose a sample of 100 items has a mean of 50 and a standard deviation of 20. Our interval estimate (with 95 per cent confidence) for the population mean is:

$$\hat{\mu} = 50 \pm 1.96 \times 20/\sqrt{100}$$

i.e.   50 ± 4 (approximately)

If this is not sufficiently precise, we could draw a larger sample and reduce the sampling error. If we wanted the sampling error to be 2 rather than 4, then:

$$2 = 1.96 \times 20/\sqrt{n}$$

so

$$2/1.96 = 20/\sqrt{n}$$

and

$$n = (20 \times 1.96/2)^2$$
$$= 384$$

If we draw another sample of 384 items then this should give us the degree of precision we require from our interval estimate. We can see, then, that if we are prepared to state in advance the size of the sampling error that would satisfy us (call this r), and then undertake a pilot sample to determine the size of the sample standard deviation (s), then the sample size n that will give the required sampling error is

$$n = (1.96s/r)^2$$

with 95 per cent confidence, or:

$$n = (2.576s/r)^2$$

with 99 per cent confidence.

We can use a similar expression to find the sample size necessary to obtain a sampling error $\pm r$ when estimating a population proportion. This is:

$$n = \frac{(1.96)^2 p(1 - p)}{r^2}$$

with 95 per cent confidence, or:

$$n = \frac{(2.576)^2 p(1 - p)}{r^2}$$

with 99 per cent confidence.

Example 2

A sales manager has conducted a survey, and found that his competitor's customers intend to switch to his product on their next purchase. He urges the board to sanction a capacity expansion. Before doing this, the board want to be 99 per cent certain that the sales manager's estimates are correct to within 0.5 per cent. What size sample should be examined to check the sales manager's claim?

$$n = \frac{(2.576)^2 \times 0.05 \ (1 - 0.05)}{(0.005)^2} = 12,608$$

Obviously, such an exceptionally large sample would cost a lot of money!

9.5 Estimating a population variance

From what you have learnt in this chapter, you might expect that an unbiased estimate of the population variance is the sample variance, but unfortunately this is not so. *The sample variance is a biased estimate, as it underestimates the population variance.* To obtain an unbiased estimate we must increase the sample variance, and we do this by applying *Bessel's correction* which is to multiply the sample variance by $n/(n - 1)$, i.e.

$$\hat{\sigma}^2 = \frac{s^2 \times n}{n - 1}$$

or what amounts to the same thing

$$\hat{\sigma}^2 = \frac{\Sigma(x - \bar{x})^2}{n - 1}$$

Now the adjustment we make to the sample variance to obtain an unbiased estimate of the population variance will depend upon the size of the sample we draw. The larger the value of n, then the smaller is the adjustment that we make. This is illustrated in the table below.

n	$n/(n - 1)$
2	2.000000
5	1.250000
10	1.111111
50	1.020408
100	1.010101
1000	1.001001

So we can see that when the sample size is two, we must increase the sample variance by 100 per cent in order to obtain an unbiased estimate of the population variance. If the sample size increases to 50 items, then the size of the adjustment falls to 2 per cent, but if we increase the sample size to 1,000 then the adjustment necessary is only 0.1 per cent.

We have applied Bessel's correction to variate sampling, but it applies equally well to attribute sampling. If the sample is large, then the variance of the sample proportion is a good approximation to the variance of the population – but if the sample size is small, then Bessel's correction must be applied, i.e.

$$\sigma^2 = \frac{p(1-p)}{n} \times \frac{n}{n-1}$$

$$= \frac{p(1-p)}{n-1}$$

Now, the importance of Bessel's correction lies in the implications it has when we either predict sample behaviour, or make an estimate of the population mean or proportion. So far, we have used Z scores from the normal distribution when we set up confidence intervals or make interval estimates, but this can only be justified as long as the population standard deviation is known or can be estimated with great precision (in the latter case if the sample is large). However, if the sample is small then the normal distribution will not apply: we will have no justification at all for using Z scores for the calculation of confidence intervals. Suppose we have a sample of n items with a mean x and a standard deviation s, then

$$\hat{\mu} = \bar{x}$$
$$\hat{\sigma}^2 = s^2 \times n/(n-1)$$

$$\hat{\sigma} = s \times \sqrt{\frac{n}{n-1}}$$

The estimated standard error of the mean will be

$$\hat{\sigma}/\sqrt{n} = s \times \sqrt{\frac{n}{n-1}} \times 1/\sqrt{n}$$

$$= s/\sqrt{(n-1)}$$

Now if the sample is large, we can make a 95 per cent interval estimate like this

$$\hat{\mu} = \bar{x} \pm 1.96 \times s/\sqrt{(n-1)}$$

But suppose the sample is small – then the constant 1.96 cannot be used. You see, the problem is that by using small samples we have introduced further errors into our estimates. In estimating the population mean, we have also had to estimate the population variance – an estimate which itself is subject to error. The error we quoted in our interval estimate, then, is too small and we should increase it; we must multiply the standard error by a constant *greater than* 1.96. Just how much more than 1.96 should our constant be depends upon the accuracy of our estimate of the population variance – which in turn depends on the sample size.

The problem of the size of the constant to use was investigated by W. S. Gosset in 1908, who published his results under the pseudonym 'A Student' (his employers would not allow him to use his real name). He realised that the standard normal Z scores would not apply to small samples, so he replaced them with a statistic he called t. Gosset discovered the sampling distribution of t, which takes into account the variations in s as well as the variations in x. For each sample size, then, there is a different t-distribution, so it is vitally important that we use the right one. Now Gosset did not tabulate the *t*-distribution against the sample size – instead he tabulated it against the number of *degrees of freedom* in the sample. Don't worry about what this means just yet – for the moment just accept that the number of degrees of freedom is the denominator in Bessel's correction, i.e. one less than the sample size.

Turn to the tables at the end of this book and you will find some values of the *t*-distribution (reasonably complete tables as we gave for the normal distribution would fill an entire book). Tabulated in the first column is the number of degrees of freedom, and in the first row are the varying probability levels. To see how to use these tables, let us consider an example.

Example 3

A sample of ten items has a mean of 35 and a standard deviation of 10. Estimate the mean of the population from which the sample was drawn.

First, let us make the point estimate

$$\hat{\mu} = \bar{x} = 35$$

To obtain the interval estimate we must estimate the standard error of the mean.

s.e. of sample mean $= s/\sqrt{(n-1)} = 10/\sqrt{(10-1)} = 3.33$

The interval estimate of the population mean is

$$\hat{\mu} = 35 \pm t \times 3.33$$

The appropriate value of *t* depends on the degrees of freedom in the sample, and the level of confidence we require for the interval estimate. In this case, the sample has 9 degrees of freedom, so the *t* value we require is given in the ninth row of the table.

Suppose we wished to set up a 95 per cent interval estimate: the *t* value we require will be in the column headed Pr = 0.025, *not* Pr = 0.05 because, like the normal distribution table, the *t* table gives the area in the right hand tail (Fig. 9.2.).

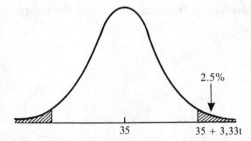

2.5%

35   35 + 3,33t

Figure 9.2

So the appropriate *t* value for this interval estimate is 2.262, and the 95 per cent interval estimate for the population mean is

$$\hat{\mu} = 35 \pm 2.262 \times 3.333$$
$$= 35 \pm 7.54$$

Likewise, the *t* value for a 99 per cent interval estimate would be found in the column headed

Pr = 0.005, i.e.

$$\hat{\mu} = 35 \pm 3.250 \times 3.333$$
$$= 35 \pm 10.833$$

As the sample size increases, the t-distribution becomes closer to the normal distribution. In fact, if the sample has 120 degrees of freedom (i.e. a sample of 121 items) then a 95 per cent interval estimate of the population mean is

$$\hat{\mu} = \bar{x} \pm 1.98 \text{ s.e.}$$

whereas for a 'large' sample it would be

$$\hat{\mu} = \bar{x} \pm 1.96 \text{ s.e.}$$

— hardly any difference. In fact, most statisticians argue that if a sample has more than 30 items, it makes little difference whether the normal distribution or the t-distribution is used. This is why statisticians call 'large' samples those with more than 30 items.

9.6 Degrees of freedom

The time has now come to examine what is meant by degrees of freedom — a concept that creates problems for most students of statistics. The concept arises because sample results vary: if we use samples to estimate (say) a population mean, then we will not obtain the same estimate from each sample. Moreover, the smaller the sample size we use, the greater will be the variation in our estimates. Suppose we draw a sample of five items from a certain population and found the mean to be 20. Of course, another, different sample of five items could also yield the same estimate, and we will now examine the values it could contain. If the sample is to contain five items and have a mean of 20, then the sample total $(\sum x)$ must be $5 \times 20 = 100$. In fact, we are perfectly 'free' to assign arbitrary values to any *four* of the five items in the sample. Suppose the first four items sampled are 10, 25, 30 and 18: then the last item *must* be $100 - (10 + 25 + 30 + 18) = 17$ — it couldn't possibly be any other value. So if we 'fix' the sample mean at x in a sample of n items, we can arbitrarily assign values to $n - 1$ items in the sample: we have $n - 1$ degrees of freedom after fixing the sample mean.

9.7 Pooled estimates of the population mean

If we draw two samples from the same population we may find that the first sample of n_1 items has a variance s_1^2. From this sample we would estimate the population variance to be

$$\frac{s_1^2 n_1}{n_1 - 1}$$

Suppose we draw a second sample of n_2 items with a variance s_2^2, the estimated population variance from this sample is

$$\frac{s_2^2 n_2}{n_2 - 1}$$

Now it would seem logical to argue that we can get a better estimate of the population variance by pooling the results of both samples than we could by either sample alone. We pool the variances like this:

$$\sigma^2 = \frac{s_1^2 n_1 + s_2^2 n_2}{n_1 + n_2 - 2}$$

In this case, we have $n_1 + n_2 - 2$ degrees of freedom associated with our estimate.

If the samples were large, then there would be no need to include the correction factor (-2) in the denominator as the difference in our results would be negligible. Suppose we draw a sample of 1,000 items with a variance of 90, and a second sample of 1,200 items with a variance of 84, then our estimate of the population variance is

$$\sigma^2 = \frac{90 \times 1,000 + 84 \times 1,200}{1,000 + 1,200 - 2} = 86.806,$$

which incorporates the adjustment, or

$$\sigma^2 = \frac{90 \times 1,000 + 84 \times 1,200}{1,000 + 1,200} = 86.72,$$

without the adjustment.

The difference in our estimates is extremely small. But if the samples were small, say 10 and 16 respectively

$$\sigma^2 = \frac{90 \times 10 + 84 \times 16}{10 + 16 - 2} = 93.5,$$

which incorporates the adjustment, or

$$\sigma^2 = \frac{90 \times 10 + 84 \times 16}{10 + 16} = 86.31,$$

without the adjustment.

The difference is alarming and serves as a warning not to ignore the adjustment when dealing with small samples.

9.8   Finite population adjustments

So far, we have been assuming that we have been drawing samples from an infinite population, or (what amounts to the same thing) if we sample from a finite population then we sample with replacement. This is not a realistic assumption; although populations under consideration may be very large, they will still be finite, and it is difficult to imagine why a statistician would practice replacement sampling. We must now examine the effects of finite populations on sampling distributions. Well, the first thing to notice is that if the population is finite and we use the formulas developed so far, then we will be *overstating* the situation. For example, if we are variate sampling, drawing a sample of n items from a population of N items, then the expression $\sigma/\sqrt{n}$ would overstate the standard error of the mean. Can you see why this must be so? With replacement sampling, then the largest and smallest items in the population can both be drawn more than once within the same sample, but with non-replacement sampling this cannot happen. The correct value for the standard error is obtained by using a correction factor: we multiply the standard error by

$$\sqrt{\frac{N-n}{n-1}}$$

This correction factor should be applied to both the standard error of the mean and the standard error of the proportion. However, in many cases it makes sense to ignore the finite population adjustment factor. If N is very large in comparison to n, then the adjustment factor will be very close to 1, and so will not be worthwhile using. For example, if we draw samples of 100 from a population of 1,000 then the size of the adjustment would be

$$\sqrt{\frac{1,000-100}{1,000-1}} = 0.949$$

but if we draw samples of 100 from a population of 100,000 then the adjustment would be

$$\sqrt{\frac{100,000-100}{100,000-1}} = 0.995,$$

which is so near to 1 that it hardly seems worth bothering with. (Remember that multiplying a number by 1 leaves the number unchanged.)

Example 4

Suppose a sample of 100 items has a mean of 50 and a standard deviation of 5. Obtain a 95 per cent interval estimate of the mean of the population from which the sample was drawn, given that there are 1,000 items in the population.

Standard error of sample mean (uncorrected) =

$\sigma/\sqrt{n} = 50/\sqrt{100} = 5$

Correction factor $= \sqrt{\dfrac{1,000-100}{1,000-1}} = 0.949$

Corrected standard error $= 5 \times 0.949 = 4.745$

95 per cent interval estimate $= 50 \pm 1.96 \times 4.745$
$= 50 \pm 9.3.$

Review questions

9.1 A simple random sample of sales invoices was taken from a very large population of sales invoices. The average value was found to be £18.50 with a standard deviation of £6. Obtain a 95 per cent interval estimate for the average value of all sales.

9.2 A population of sales invoices is to be sampled so that the mean value per sale can be estimated. One hundred sales invoices are selected at random from this population and classified by value, as shown in the following frequency distribution.

Value (£)	No. of sales invoices
0–50	10
50–100	9
100–150	15
150–200	22
200–250	13
250–300	7
300–350	9
350–400	5
400–450	6
500–550	4

Use the sample information to construct interval estimates for the population mean at the 95 per cent and 99 per cent levels of confidence.

9.3 A manufacturer wishes to estimate the mean weight of sacks of carbon black. He would be satisfied if his estimate was within ±5 kg of the true mean weight and be 99 per cent sure of his estimate. An initial sample gives a standard deviation of 15 kg. What size sample yields the required estimate?

9.4 A local election is being held and there are two candidates, Smith and Jones, for one vacancy. A random sample of the electorate reveals that 55 per cent of them will vote for Smith. How confident can you be that Smith will be elected, if the sample size was 100? How large a sample would be needed for the 55 per cent sample results for Smith to make you 99 per cent confident of a win for him?

9.5 In a random sample of 200 garages it was found that 79 sold car batteries at prices below that recommended by the manufacturer.

(a) Estimate the proportion of all garages selling below the list price.

(b) Calculate 95 per cent and 99 per cent confidence limits for this estimate and explain what these mean.

(c) What size sample would have to be taken in order to be 95 per cent certain that the population proportion could be estimated within 2 per cent?

9.6 Nine bags of sugar were selected and carefully weighed. The mean weight was 1.004 kg and the standard deviation 0.0002 kg. Find the 95 per cent interval estimate for the mean weight of all bags.

9.7 A sample of 10 items has a mean 20 and a standard deviation 0.5. A second sample of 15 items has a mean 19.5 and a standard deviation 0.4. Assuming that both samples came from the same population, estimate the population mean and standard deviation.

9.8 A civil engineering contractor buys a job lot of 10,000 pipes which have a nominal length of 10 metres each. A random sample of 100 pipes were chosen and their lengths carefully measured. The mean length was 9.59 metres and the standard deviation 0.01 metres. Give a 95 per cent interval estimate for the total length of piping in the job lot.

9.9 A by-election has been declared at Anytown, which has 40,000 voters. A random sample of 100 voters reported that 53 per cent can be expected to vote Democrat. Assuming that 70 per cent of the registered voters do indeed vote, give a 95 per cent interval estimate for the proportion voting Democrat at the election.

10 Hypothesis testing

10.1 Introduction

In the previous chapter, we used sampling as a basis for estimation. We shall now turn our attention to using sampling as a basis for decision taking. Now as you already know, we cannot be completely sure of sample results, so if we base decisions on sampling evidence, it is quite conceivable that we make a wrong decision. It is a major task of statistics to evaluate the probability of making the wrong decision, and to keep this probability to a minimum.

10.2 Type 1 errors

We will now examine a problem that faces many manufacturers: how to control the quality of output. Let us suppose that a manufacturer produces woodscrews by a completely automatic process. One of the problems with the automatic process is that occasionally a screw misses the threading stage, so such screws should be rejected. Now there is little that can be done about this, and the manufacturer has installed machinery that would automatically reject such screws. From past experience it is known that about 5 per cent of output would be rejected by this machine. Now although there is not much chance of bettering this rate, machine malfunction can cause the rate to worsen considerably. The problem experienced so far is that it has been difficult to detect just when the rate of rejects has increased because the quantity of output is so great. Obviously, the manufacturer wants to keep the rate as near to 5 per cent as possible – after all, rejects cost money! What he decides to do is to install an automatic sampling device which randomly selects 1,000 screws, sorts them, and counts the number of rejects.

The purpose of drawing these samples is, of course, to detect when the reject rate has worsened, and do something about it. Taking remedial action involves stopping the process, and this will be expensive as output will be lost. So if remedial action is taken, the manufacturer wants to be pretty sure that the reject rate has worsened. What he wants is some *decision rule* to determine when to stop the process and make the necessary adjustments. Now as 5 per cent of the output is rejected, the manufacturer can expect, on average, 50 rejects in each sample. Suppose he errs on the cautious side and adopts the decision rule – 'stop the process if the number of defectives exceeds 60'.

Unfortunately, this rule does not exclude the possibility of taking the wrong decision; i.e. stopping the process when the reject rate has not increased. Using our knowledge of the binomial distribution, we can calculate the probability of making the wrong decision. It will be the wrong decision if the number of defectives exceeds 60 within the sample, but the overall reject rate is still 5 per cent. The appropriate binomial distribution is $(0.95 + 0.05)^{1,000}$ and the probability we require is

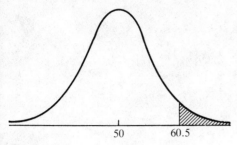

Figure 10.1

$P_{(61)} + P_{(62)} + P_{(63)} + \cdots + P_{(1,000)}$. Now obviously it would be advisable to use the normal approximation to the binomial distribution. First, we require the standard deviation of this distribution.

$$\sigma = \sqrt{npq}$$
$$n = 1{,}000, \qquad p = 0.05, \qquad q = 0.95$$
$$\sigma = \sqrt{1{,}000 \times 0.05 \times 0.95}$$
$$= 6.89$$

The probability that the sample contains more than 60 rejects is shown in the shaded area of Fig. 10.1 (remember that we must make the 0.5 adjustment when using the normal approximation to the binomial distribution). The Z score is:

$$\frac{60.5 - 50}{6.89} = 1.52$$

Consulting the normal distribution tables, the probability we require is 6.43 per cent. So if the manufacturer decides to stop the process if the number of rejects in the sample exceeds 60 on the assumption that the overall reject rate has risen above 5 per cent, then the probability that he has made the wrong decision is 6.43 per cent.

Is this an acceptable risk? Well, this is really a decision that only the manufacturer can take. To decide whether the risk is too high, he must weigh up what is at stake. Can you see how the manufacturer can reduce the chance of taking the wrong decision? All that he need do is to to alter his decision rule, taking a number of rejects greater than 60 as the borderline. If, for example, his decision rule was to stop the process if the number of rejects in the sample exceeds 65, then the Z score would be

$$\frac{60.5 - 50}{6.89} = 2.25$$

and the probability of taking the wrong decision would be 1.22 per cent.

The relationship between the probability of taking the wrong decision and the number of rejects in the decision rule is something we will examine in more detail later. What we shall now do is to express the problem in statistical jargon. The manufacturer formulates the *hypothesis* that the overall number of rejects is 5 per cent. He uses the following decision rule: if the number of rejects in the sample exceeds 60, reject the hypothesis and stop the process, otherwise accept the hypothesis. Notice that we have not altered the philosophy of his decision rule − all we have done is to introduce the word 'hypothesis'. Now we have seen that there is a danger of making the wrong decision and needlessly stopping the process. Using statistical jargon, there is a danger of *rejecting the hypothesis when it should be accepted*. Statisticians refer to this risk as a *Type 1 error*. We have seen that the risk of a Type 1 error will vary according to the number of rejects quoted in the decision rule. We

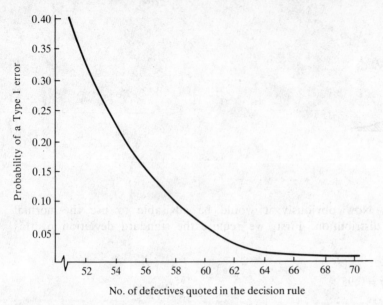

Figure 10.2

can quantify this risk by calculating the Z score for each of the rejects quoted in the decision rule.

Number of rejects quoted in the decision rule	Z score	Probability of a Type 1 error
52	0.36	0.3594
54	0.65	0.2578
56	0.94	0.1763
58	1.23	0.1093
60	1.52	0.0643
62	1.81	0.0351
64	2.10	0.0179
66	2.39	0.0084
68	2.69	0.0036
70	2.98	0.0014

It would be useful to draw a graph of this data and this is done in Fig. 10.2. The manufacturer can see at a glance the risk of incurring a Type 1 error as he varies the number of rejects in his decision rule. Of course, it is quite possible to state the degree of risk of a Type 1 error that would be acceptable, and read off from the graph the number of rejects in the decision rule that would yield this degree of risk. Suppose, for example, he would consider acceptable a 5 per cent chance of a Type 1 error, then consulting the graph we see that the decision rule should be: 'stop the process if the number of rejects exceeds 61'. Using statistical jargon, we say that the graph is a *performance chart* for Type 1 errors in our decision rule. Notice that as we increase the number of rejects in our decision rule, the probability of a Type 1 error decreases.

10.3 Type 2 errors

Let us summarise what we have done so far. We have formulated the hypothesis that the number of rejects is 5 per cent, and we use the decision rule that if the number of rejects in the sample exceeds a certain number n, then we would reject the hypothesis and stop the process, otherwise we accept the hypothesis and allow the process to continue. We have seen that in operating this decision rule there is a danger of rejecting the hypothesis when it should have been accepted – so we would needlessly stop the process. This is because it may be possible that the number of rejects in the sample exceeds n even though the overall number of rejects is still 5 per cent. We call this a Type 1 error, and the performance chart shows how the risk of a Type 1 error varies with the size of n in our decision rule.

Now the Type 1 error is not the only risk the manufacturer runs when operating the decision rule. Originally, he operated the: 'stop the process if the number of rejects in the sample exceeds 60, otherwise allow the process to continue'. The philosophy underlying this rule is that if the number of rejects is 60 or less, then the overall rate of rejects could still be 5 per cent. The fact that the number of rejects exceeds the expected 50 can be explained by sampling error. Now it is quite possible for the overall rate of rejects to rise above 5 per cent, but the sample drawn will contain 60 or less rejects. Should this be the case, then he will be allowing the process to continue when really he should stop the process and make the necessary adjustments. Returning to statistical jargon, there is a danger of *accepting a hypothesis that should be rejected*. Such risks are referred to as *Type 2 errors*, and we should be able to calculate the probability of such errors

Suppose the overall rate of rejects rose to 7 per cent. For samples of 1,000, we have $n = 1,000$, $p = 0.07$ and $q = 0.93$. So the appropriate binomial distribution is:

$$(0.93 + 0.07)^{1,000}$$

mean = $np = 1,000 \times 0.07 = 70$
standard deviation = $\sqrt{npq} = \sqrt{1,000 \times 0.07 \times 0.93} = 8.07$

We can use the normal approximation to the binomial distribution to obtain the probability of obtaining 60 or less rejects in a sample of 1,000 items (Fig. 10.3).

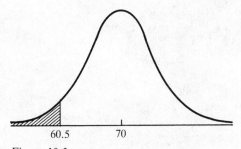

60.5   70

Figure 10.3

The Z score is

$$\frac{60.5 - 70}{8.07} = -1.18$$

and consulting the normal tables we see that the probability of a Type 2 error is 0.119. In other words, there is a 11.9 per cent chance that the decision rule will fail to detect a rise in the overall rate of rejects to 7 per cent. So there is a 11.9 per cent chance that the manufacturer will make the wrong decision and allow the process to continue.

The probability of making a Type 2 error will depend on the actual rate of rejects produced by the process. If we list some possible reject rates, we can calculate the appropriate probabilities of Type 2 errors.

Proportion of rejects produced by the process	Mean	Standard deviation	Z score	Probability of Type 2 error
0.055	55	7.21	0.76	0.7764
0.060	60	7.51	0.07	0.5279
0.065	65	7.80	−0.58	0.2810
0.070	70	8.07	−1.18	0.1190
0.075	75	8.33	−1.74	0.0409
0.080	80	8.58	−2.27	0.0116
0.085	85	8.82	−2.78	0.0027
0.090	90	9.05	−3.26	0.0004
0.095	95	9.27	−3.72	0.0001
0.100	100	9.49	−4.16	0.0000

Figure 10.4 shows this information graphically. Notice that as the proportion of rejects produced by the process increases, the smaller is the probability of making a Type 2 error. Notice also that to calculate the probability of making a Type 2 error we must make an assumption about the value of the overall rate of rejects produced by the process. For a Type 1 error this was not necessary, so it is easier to calculate the performance chart for Type 1 errors. Also, the performance chart for Type 2 errors is not in a very convenient form – it does not show the manufacturer how he can lessen the chance of a Type 2 error.

Earlier, we calculated that if the overall rate of rejects produced by the process was 7 per cent, then, given our decision rule, the probability of a Type 2 error is 11.9 per cent. Suppose we changed

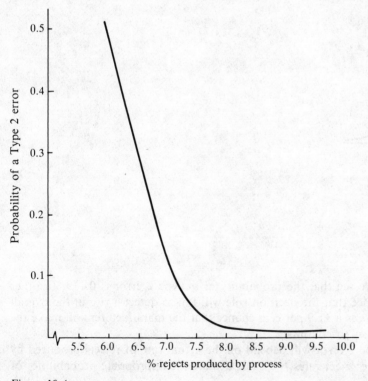

Figure 10.4

the decision rule to: 'stop the process if the number of rejects exceeds 55'. Now if the overall rate of rejects is 7 per cent and the sample contains 55 rejects or less, then the process will not be stopped when it should be, and the probability of this happening can again be calculated.

$$Z = \frac{55.5 - 70}{8.07} = -1.80$$

and from the normal tables, the probability of taking the wrong decision is 3.59 per cent. So if we reduce the number of rejects in our decision rule, this decreases the probability of a Type 2 error. Unfortunately, this action would automatically increase the probability of a Type 1 error.

Summarising, then, a Type 1 error occurs when we reject a hypothesis that should be accepted, and a Type 2 error occurs when we accept a hypothesis that should be rejected. Many statisticians summarise the distinction in a table like this:

		Decision	
		Accept	Reject
	True	Correct decision	Type 1 error
Hypothesis is	False	Type 2 error	Correct decision

If we state our hypothesis and decision rule, we can immediately calculate the probability of a Type 1 error, but we cannot calculate the probability of a Type 2 error without introducing other assumptions. In other words, to calculate the probability of Type 1 errors, we keep the hypothesis constant and vary the decision rule; but to calculate the probability of Type 2 errors, we keep the decision rule constant and vary the hypothesis. We can reduce the probability of a Type 1 error by increasing the number of rejects in our decision rule, but unfortunately this would increase the probability of a Type 2 error. Now this is quite a problem, and we will try to reconcile it in the next section.

10.4 Significance testing

In the last section, we found no difficulty at all in calculating the probability of a Type 1 error. The reason for this was the form taken by our hypothesis: that the overall rate of rejects was 5 per cent. So we could put $p = 0.05$ into the binomial distribution and calculate the probability of a Type 1 error with ease. Notice that in applying this hypothesis, we are assuming that the reject rate found in the past has continued, i.e. that the reject rate has not changed. We have done this despite the fact that the purpose of our investigation is to detect when the overall reject rate has worsened. Because our hypothesis has assumed that 'nothing has happened', it is called a *null hypothesis* and we use the symbol H_0 to stand for a null hypothesis. Now you may object that this seems rather strange – the purpose of sampling is to detect when the reject rate has worsened, so we should be testing the hypothesis that the reject rate is greater than 5 per cent. Now this is all very well, but it will mean that we are unable to calculate the probability of a Type 1 error without stating by *how much* the reject rate has worsened. In other words, stating that $p > 0.05$ means that we do not have a value for p to insert into the binomial distribution, and so we will be unable to calculate the probability of a Type 1 error. It is for this reason that we will ensure that from now on all the hypotheses we test will be null hypotheses.

The testing of null hypotheses has two implications which we must now examine. Firstly, the null hypothesis involves stating the complete opposite to what you are trying to detect. In the example we have been considering so far, we were trying to detect whether the reject rate had worsened, but the null hypothesis is that the reject rate is still 5 per cent. Again, suppose we were trying to detect for bias in a coin – the appropriate null hypothesis would be that the coin is unbiased. It would follow from this that the probability of a head is 0.5, so we could then calculate the probability of a Type 1 error. We can now state a rule for formulating null hypotheses – if you are trying to detect change, then the appropriate null hypothesis is that no change has occurred.

The second implication of formulating a null hypothesis is that it is of no help whatsoever in calculating the probability of a Type 2 error. To do this, we must state an *alternative hypothesis*. This is precisely what we did when calculating the probabilities of Type 2 errors in the last section. In fact, we stated a number of alternative hypotheses ($p = 0.055$, $p = 0.06$, $p = 0.065$ etc.). Now obviously there is a problem here: if we reject the null hypothesis, then it would seem reasonable to accept the alternative hypothesis: but which alternative hypothesis should we accept?

Let us now see if we can find a way round this problem. We incur Type 2 errors if we accept a hypothesis that should be rejected – so if we never accept a hypothesis we cannot make a Type 2 error! Let us restate the problem we have been considering so far, so as to investigate this process more closely. We know from past experience that the process is, on average, rejecting 5 per cent of all screws produced. Our null hypothesis is that the overall rate of rejects produced is 5 per cent, and we operate the decision rule 'reject the null hypothesis if the number of rejects in the sample exceeds 60, otherwise *reserve judgement*'. Notice that we never accept the null hypothesis, and in many statistics textbooks you will come across the conclusion 'no need to reject the null hypothesis' – which of course doesn't mean that it should be accepted! The philosophy behind this argument is that although we would expect samples to contain 50 rejects on average, individual sample results will vary. However, samples that contains 60 or less rejects do not show a *significant* departure from the expected 50 rejects. In other words, such results can be attributed to chance or sampling fluctuations. However, samples with more than 60 rejects cannot be attributed to chance, the *sample result is significant* and we must reject the null hypothesis. Notice the use of of the word 'significant' – many statisticians give the title *significance testing* to the methods we have been describing.

So you see, it is possible to avoid making Type 2 errors altogether. However, you may complain that the reasoning behind this is very suspect. Initially, we were trying to detect when the overall reject rate rose so that we could do something about it. Our null hypothesis was that the overall reject rate was 5 per cent, and our decision rule was 'stop the process if the number of rejects exceeds 60', i.e. reject the null hypothesis. Now suppose that the number of rejects was 60 or less, and that this would induce us to conclude that there is no reason to reject the null hypothesis. Now think carefully about this! If the number of rejects in the sample exceeds 60, we stop the process; if the number does not exceed 60 we reserve judgement. But can we reserve judgement? If we do this, we will allow the process to continue – which is tantamount to accepting the null hypothesis! In such circumstances, then, we cannot avoid the risk of a Type 2 error!

Let us now see if we can devise an acceptable methodology for significance testing. First, we will formulate a null hypothesis (H_0) in such a way that we can calculate the probability of a Type 1 error. Also, we will formulate an alternative hypothesis (H_1) so that the rejection of H_0 automatically involves the acceptance of H_1. We will then specify the probability of a Type 1 error that we would be prepared to consider an acceptable risk. We call this probability the *level of significance*, and it is usual to specify probabilities of 5 per cent and 1 per cent. Whenever possible we word the alternative hypothesis in such a way that if we do not reject the null hypothesis we reserve judgement. If this is not possible, then the wording of the alternative hypothesis is such that the alternative to rejecting the null hypothesis is accepting it. In this latter case, we keep our fingers crossed that the probability of a Type 2 error is not too large. If Type 2 errors would be really serious, then we must examine the performance chart to see how likely they are.

10.5 One-sided and two-sided tests

We will now apply this methodology to the problem we have so far been considering. Firstly, we specify the hypotheses

H_0: $\pi = 0.05$, i.e. the overall reject rate is 5 per cent
H_1: $\pi > 0.05$, i.e. the overall reject rate is greater than 5 per cent
Level of significance = 0.05, i.e. the risk of a Type 1 error is 5 per cent

Consulting the performance chart for Type 1 errors, we see that for a 5 per cent probability we must quote 61 rejects in out decision rule, so our decision rule is 'reject H_0 if the number of rejects in the sample exceeds 61, otherwise accept H_0'. We could represent this situation diagrammatically as shown in Fig. 10.5.

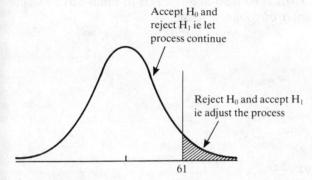

Accept H_0 and reject H_1 ie let process continue

Reject H_0 and accept H_1 ie adjust the process

61

Figure 10.5

Summarising this situation, then, we can say that on the basis of our null hypothesis we expect, on average, 50 rejects in a sample of 1,000. If the number of rejects is 61 or less, then we can say that the deviation from 50 can be attributed to chance. But if the number of rejects exceeds 61 we cannot attribute this to chance, i.e the results are significant, and we reject the null hypothesis, concluding that the reject rate is greater that 5 per cent. We realise that in using this decision rule we run a 5 per cent risk of making a Type 1 error.

In this example, we are interested in detecting shifts in the reject rate in one direction only. If the reject rate increases, then we must take some action. But if it decreases – so much the better. Tests to detect changes in one direction only are called *one-sided tests or one-tail tests*, and a glance at Fig. 10.5 will explain why this is so. However, there will be many cases where it will be more meaningful to detect changes in either direction. To take an example, suppose we know that 40 per cent of the electorate vote Socialist, and we wish to detect changes in the Socialist vote by sampling. It will be of equal interest to know whether the Socialist vote has increased or decreased. The appropriate hypotheses would be

$H_0 : \pi = 0.4$ 　　$H_1 : \pi <> 0.4$

Notice that in this case the alternative hypothesis is stated in a 'not equal to' form. Without going into the mechanics of this problem (we will do this in the next chapter) suppose the decision rule was: 'in a sample of 1,000, if the number voting Socialist is between 370 and 430 accept H_0 and reject H_1, otherwise a accept H_1 and reject H_0'. Figure 10.6 illustrates this situation.

Tests such as this are called *two-sided tests or two-tail tests*, again for obvious reasons. So if you wished to test for a change in an *unspecified direction* you should use a two-sided test, but if you

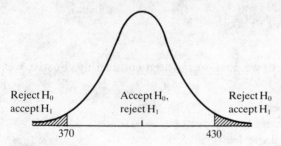

Figure 10.6

wish to test for a change in a *specified direction* you would use a one-sided test. The type of test used will be reflected in the formation of the alternative hypothesis. Which type of test is appropriate will depend upon the problem under consideration. There is no hard-and-fast rule to follow: as we shall see in the next chapter the choice is really a matter of common sense.

Review questions

10.1 Suppose a person is to be tried for a certain crime.

    (a) What is the appropriate null hypothesis?
    (b) What is the appropriate alternative hypothesis?
    (c) What is the consequence of a Type 1 error?
    (d) What is the consequence of a Type 2 error?
    (e) Which type of error is it more important to minimise?

10.2 From past experience, it has been found that 5 per cent of a certain output is rejected. To test the null hypothesis H_0: $\pi = 0.05$, a sample of 400 items is drawn. Form the decision rules for testing H_0 at the 1 per cent level if

    (a) the alternative hypothesis is one-sided,
    (b) the alternative hypothesis is two-sided.

10.3 Suppose that in the previous question we adopt the one-sided alternative hypothesis, and that the reject rate rises to 7 per cent. What is the probability of a Type 2 error?

11 Statistical significance

11.1 Introduction

In the previous chapter, we indicated how statistics could be used as an aid to decision taking. The methodology we used was as follows. We formulated two hypotheses: a null hypothesis and an alternative hypothesis, in such a way that rejecting one meant accepting the other. The wording we chose for the null hypothesis enabled us to calculate the probability of a Type 1 error. We then stated a decision rule, for example 'reject the null hypothesis and accept the alternative hypothesis if the number of rejects in the sample exceeds x'. If we rejected the null hypothesis, we then calculated the probability of a Type 1 error, and we called this probability the *level of significance*. This is the methodology of the statistical technique called *significance testing*. In this chapter, we will move from the methodology to the technique itself. In particular, we will be dealing with tests involving sample means and sample proportions. Before we do this, however, we must examine the choice of an appropriate level of significance.

11.2 Tests of significance

As we saw in the previous chapter, when we altered our decision rule in such a way that we reject the null hypothesis only if the number of rejects is greater than previously stated, then we reduced the probability of a Type 1 error. We must now decide on the risk of a Type 1 error that we could consider acceptable. Now of course, the level of acceptable risk is a matter of personal choice, and will vary with the problem we are considering. We would want, for example, a lower level of risk if we are testing for efficacy of a new drug, than we would want if we were testing a marksman for bias in shooting. Despite this essentially subjective nature of the level of risk, statisticians do apply conventions. They consider a 5 per cent probability of a Type 1 error as being acceptable, and construct their decision rules accordingly. If the hypothesis is rejected on the basis of this decision rule, we state that the sample result is *significant at the 5 per cent level*. In other words, we do not think that the sample results can be attributed to chance, and our probability of making a Type 1 error is 5 per cent. Also, it is useful to construct a second decision rule based on the probability of a Type 1 error being 1 per cent, so if the null hypothesis is rejected on the basis of this rule, we state that the result is *significant at the 1 per cent level*. So applying our decision rules to sample results, there are four possible conclusions we could reach:

   (a) Accept the null hypothesis and reject the alternative hypothesis. In this case, we are saying that the sample results can be attributed to chance. However, as we saw in the last chapter, reaching this conclusion leaves us wide open to Type 2 errors, and the only way to avoid them is to conclude that

(b) there is insufficient reason to reject the null hypothesis – a conclusion that does not conclude anything and suggests the necessity for further sampling. However, if anything useful is to be achieved, then sooner or later we must risk the possibility of a Type 2 error and come to some definite conclusion.

(c) Reject the null hypothesis and accept the alternative hypothesis at the 5 per cent level, but not at the 1 per cent level. Here we are stating that the occurrence of the sample results cannot be attributed to chance, and that the probability of a Type 1 error is less than 5 per cent but greater than 1 per cent.

(d) Reject the null hypothesis (and so accept the alternative hypothesis) at the 1 per cent level. In this situation we are taking a less than 1 per cent chance of incurring a Type 1 error.

Now let us look at the actual wording of the decision rule. We have implied that the decision rule will be stated in terms of the sample statistic, for example 'reject the null hypothesis if the proportion rejected exceeds x'. How can we find the value of x? To do this, we must state the level of significance (which we will assume to be 5 per cent), the sample size (call this n) and the null hypothesis about the population proportion (call this π). The value of x, then, is the proportion that would be exceeded in our samples on only 5 per cent of occasions. Using the normal distribution tables, we can conclude that

$$x = \pi + 1.645 \sqrt{\frac{\pi(1 - \pi)}{n}}$$

Alternatively, we may be testing for a change in the population proportion – a movement in either direction. Here, a two-sided decision rule would be appropriate – for example, 'accept the null hypothesis if the proportion defective in the sample is outside the range x_1 to x_2, otherwise reject it'. Keeping the same level of significance, this range can be obtained from the 95 per cent confidence interval

$$\pi \pm 1.96 \times \sqrt{\frac{\pi(1 - \pi)}{n}}$$

So we can see here that the decision rule is merely a confidence interval. In fact, if you think about it, you will realise that a confidence interval is a prediction about a sample statistic on the assumption that the null hypothesis is true.

Now although there is absolutely nothing wrong with using decision rules to test hypotheses, statisticians do not in fact use them. Taking a one-sided test, a statistician would argue something like this: the null hypothesis would be rejected if

$$p > \pi + 1.645 \times \sqrt{\frac{\pi(1 - \pi)}{n}}$$

where p is the sample proportion. Rearranging this condition, the null hypothesis would be rejected if

$$\frac{p - \pi}{\sqrt{\frac{\pi(1 - \pi)}{n}}} > 1.645$$

You should recognise the left-hand side of this condition: it is the Z score of the sample proportion, and because it is used to test the null hypothesis we call the Z score a *test statistic*. The right-hand side of the condition is the value for the Z score that would be exceeded only on 5 per cent of occasions: if it is exceeded by the test statistic then we would reject the null hypothesis. So the right-hand side of this expression is a kind of *critical value* for our test statistic.

This, then is how we shall undertake significance testing. We will set up a null hypothesis and an alternative hypothesis. If we assume that the null hypothesis is true, then we can calculate a test statistic (which basically is the standard score of our sample statistic) and compare the test statistic with its critical value. If the test statistic is greater than its critical value, then we reject the null hypothesis and accept the alternative hypothesis. However, if the test statistic is less than its critical value, we would accept the null hypothesis and reject the alternative hypothesis.

11.3 Test of population means using large samples

When we discussed sampling theory, we saw that if we drew large samples from a population with a known mean and standard deviation, then the sample means will be normally distributed. So, if we use large samples to test hypotheses about a population mean, then the test statistic would be the Z score of the sample mean, i.e.

$$\text{Test statistic } Z = \frac{|\bar{x} - \mu|}{\text{standard error}}$$

Notice that the test statistic is always taken as positive. We can use our knowledge of confidence intervals to state the critical values of Z. The critical value will depend upon the level of significance and on whether the test is one-sided or two-sided.

	Critical values for Z	
	5% level	1% level
One-sided test	1.645	2.323
Two-sided test	1.96	2.576

Can you see why we always take the absolute value of the test statistic? If we did not, then we would have to state *two* critical values for each level of significance for a two-sided test – for example at the 5 per cent level the critical values would be 'greater than 1.96' and 'less than −1.96'. Now as, say, a test statistic of 2 and −2 are equally significant, it makes sense to take the absolute value of the Z score.

Example 1

A sugar refiner packs sugar into bags weighing, on average, 1 kilogram. Now the setting of the machine tends to 'drift', i.e. the average weight of the bags filled by the machine sometimes increases and sometimes decreases. If the mean weight of bags increases, then in effect he is actually giving away some sugar. If the mean weight decreases then he might find himself in trouble with the Weights and Measures Inspector. So it really is important that he controls the average weight of bags of sugar. He wishes to detect shifts in the mean weight as quickly as possible, and reset the machine. In order to detect shifts in the mean weight, he will periodically select a sample of 100 bags, weigh them and calculate the sample mean.

In effect, the refiner is performing a significance test, adopting as his null hypothesis that the population mean is 1 kilogram, i.e.

$H_0: \mu = 1$

Now shifts in μ in either direction are important, so the test is two-sided and the alternative hypothesis is

$H_1: \mu <> 1$

What other information do we need to know? Clearly, we need to know the variance of the weights of bags of sugar – suppose this is 0.1 kg. We can now calculate the standard error of the mean for samples of 100 bags:

s.e. $= \sigma/\sqrt{n} = 0.1/\sqrt{100} = 0.01$ kg

and if we draw a sample of 100 bags and use the sample mean $\bar{x}$ to test the null hypothesis, then the test statistic Z would be

$$\frac{|\bar{x} - \mu|}{\text{s.e.}} = \frac{|x - 1|}{0.01}$$

As the test is two-sided, we would compare the test statistic with the critical values 1.96 and 2.576.
  Suppose the sample mean was 1.03 kg, then

test statistic $Z = \dfrac{|1.03 - 1|}{0.01} = 3$

and as this exceeds the critical value we would reject the null hypothesis and accept the alternative hypothesis at the 5 per cent level of significance. We would conclude that the overall mean weight was no longer 1 kg, and run a less than 1 per cent chance of a Type 1 error.

Example 2

A certain highway in a city centre has been investigated in the past for noise pollution by vehicles. As a result of many measurements it has been found that, between 16.30 and 17.30 on any weekday, the average noise level is 130 decibels, with a standard deviation of 20 decibels. The residents are convinced that the noise level is getting worse, and having taken 50 readings, obtained an average of 134 decibels. Is their claim justified?
  Clearly, this is a one-sided test, so we can formulate

$H_0: \mu = 130$
$H_1: \mu > 130$

Test statistic $Z = \dfrac{|\bar{x} - \mu|}{\text{s.e.}} = \dfrac{|134 - 130|}{20/\sqrt{50}} = 1.41$

As this is a one-sided test, the critical values for Z are 1.645 (at the 5 per cent level) and 2.326 (at the 1 per cent level). So we see that the difference between the population mean and the sample mean is not significant – it can be accounted for by sampling fluctuations. We would accept the null hypothesis and reject the alternative hypothesis, and so cannot conclude that noise levels have increased.

11.4 Testing for a difference between means

The two examples we have considered so far assume that we have prior knowledge of the population –

in particular they assume that the population mean and standard deviation are known. Had we not known the population standard deviation, we could have estimated it using s, the sample standard deviation. No adjustment to s^2 would have been necessary because in both cases we had large samples. Now of course, it will not always be the case that we know the population mean, and if we cannot compare a sample with a population, then we must compare a sample with a sample. Tests of this nature come under the general heading of the *significance of the difference between the means of two samples*. Let us see how this test works by considering an example.

Example 3

Two statisticians (who are also football supporters) are having an argument about which team plays the most attractive, attacking football. One follows Liverpool F.C. and the other follows Manchester United. To settle the argument once and for all, they agree to apply statistical analysis. Suppose that someone had watched every game ever played by Liverpool and counted the number of attacking moves made in each game. From this data it would be possible to calculate the mean number of attacking moves per match. Now as every game has been examined, this is the *population mean*, which we shall call μ_1. Likewise, we shall call μ_2 the (population) mean number of attacking moves made by Manchester United. The appropriate null hypothesis is that there is no difference between the average number of attacking moves made by each team. In other words,

H_0: $\mu_1 = \mu_2$ or
H_0: $\mu_1 - \mu_2 = 0$

Likewise, the alternative hypothesis is that there is a difference in the mean number of attaching moves.

H_1: $\mu_1 < \, > \mu_2$

So we are performing a two-sided test. Now in practice, it is highly unlikely that the values of μ_1 and μ_2 will be known, and they will have to be estimated by sampling. Suppose that sampling the matches played gives the following results

| | Attacking moves | | Number of matches |
	Mean	Standard deviation	in the sample
Liverpool	$32 = \bar{x}_1$	$5 = s_1$	$60 = n_1$
Manchester United	$30 = \bar{x}_2$	$6 = s_2$	$50 = n_2$

The standard error of the mean for Liverpool is

$s_1/\sqrt{n_1} = 5/\sqrt{60} = 0.645$

and the standard error of the mean for Manchester United is

$s_2/\sqrt{n_2} = 6/\sqrt{50} = 0.849$

In making these calculations, we assume that the sample results for Liverpool come from a population with a standard deviation of 5 and that the results for Manchester United come from a population with a standard deviation of 6. This assumption is reasonable: the samples are large, so the sample variances are legitimate estimates of the population variances. We can now use the sample difference theorem developed earlier to calculate the standard error of the difference between the means, i.e.

$$\text{standard error of the difference} = \sqrt{\left(\frac{s_1}{\sqrt{n_1}}\right)^2 + \left(\frac{s_2}{\sqrt{n_2}}\right)^2}$$

$$= \sqrt{(0.645)^2 + (0.849)^2}$$

$$= 1.066$$

Now if the null hypothesis is true, and there is no difference between the population means, then this is tantamount to assuming that both samples could have come from the same population. Now the actual difference between the sample means is 2, but if the null hypothesis is correct then this difference does not deviate significantly from zero. In other words, we consider the difference of 2 as coming from a population of sample mean differences with a mean of zero and a standard error of 1.066. We now have all the necessary information to undertake a significance test:

Test statistic $Z = (2 - 0)/1.066 = 1.876$

As we are conducting a two-sided test, the critical values for Z are 1.96 (at the 5 per cent level) and 2.576 (at the 1 per cent level). So we must accept the null hypothesis: the samples do not provide evidence that there is a difference in the mean number of attacking moves made by the two teams. It is important to realise that in using this method of significance testing, we are assuming that the samples are *independent*, i.e. both samples are assumed to have come from the same population, and the selection of the first sample in no way affects the selection of the second. This method could not be used in 'before and after' type of experiments, i.e. when the same sample is considered under different conditions. Later, you will be shown how to deal with samples that are not independent (statisticians call them *paired samples*).

In this section, we introduced two types of significance tests and it would be useful to introduce general expressions for the test statistic. In Examples 1 and 2, we tested an hypothesis about a population mean μ, using a sample mean $\bar{x}$.

The test statistic in this case was

$$\frac{|\bar{x} - \mu|\sqrt{n}}{\sigma}$$

In Example 3, we tested for a difference in population means μ_1 and μ_2, using sample means $\bar{x}_1$ and $\bar{x}_2$, and sample standard deviations s_1 and s_2. The test statistic in this case was

$$\frac{|\bar{x}_1 - \bar{x}_2| - 0}{\sqrt{\dfrac{s_1^2}{n_1} + \dfrac{s_2^2}{n_2}}}$$

11.5   Tests of population means using small samples

Now suppose that we use small samples to test hypotheses about population means. Now if we use the methods of the last section, we will run into trouble on two accounts. First, our small sample will underestimate the population variance, so our test statistic will be wrong. Second, the means of small samples are not normally distributed, so our critical values will be wrong. We have learnt that the means of small samples have a t-distribution, and the appropriate t-distribution will depend on the number of degrees of freedom in estimating the population variance. If we use a large sample to test a hypothesis, then the critical values we use will depend upon the type of test (i.e. one-sided or two-sided). But if we use small samples, then the critical values will depend on the degrees of freedom in the sample as well as the type of test. Fortunately, we already know how to use tables of the t-distribution. Using small samples, then, both the test statistic and the critical values will be t-scores.

Example 4

The expected lifetime of electric light bulbs produced by a given process was 1,500 hours. To test a new batch a sample of 10 bulbs was taken, and showed a mean lifetime of 1,410 hours. The standard deviation is 90 hours. Test the hypothesis that the mean lifetime of the electric light bulbs has not changed, using a level of significance of (a) 5 per cent and (b) 1 per cent.

This question asks us to test that the mean has not changed, so we must employ a two-sided test.

H_0: $\mu = 1,500$
H_1: $\mu < > 1,500$

In this example, we do not know the population standard deviation, so we must estimate it using the sample standard deviation. In the previous chapter we saw that

$$\sigma^2 = \frac{ns^2}{n-1}$$

so the estimated population variance is

$$\frac{10 \times 90^2}{9} = 9,000$$

The estimated population standard deviation is

$$\sqrt{9,000} = 94.87$$

So the test statistic for our sample mean is

$$t = \frac{|1,410 - 1,500|\sqrt{10}}{94.87} = 2.999 = 3$$

Now for the critical value of t, we enter the t-table with $10 - 1 = 9$ degrees of freedom, remembering that this is a two-sided test. The critical values are 2.262 (at the 5 per cent level) and 3.25 (at the 1 per cent level). Our test statistic is significant at the 5 per cent level but not at the 1 per cent level. We reject H_0 and accept H_1, and conclude that there is some evidence to suggest that the mean lifetime of all bulbs has changed.

In fact, we could find the test statistic directly without first estimating the population variance if we use the expression

Test statistic $t = \dfrac{|\bar{x} - \mu|\sqrt{(n-1)}}{s}$

In the last example this would give

$$t = \frac{|1,410 - 1,500|\sqrt{(10 - 1)}}{90} = 3, \text{ as before.}$$

Example 5

After treatment with a standard fertiliser, the average yield per hectare is 4.2 tonnes of wheat. A super fertiliser is developed and administered to 10 hectares. The yields were 4.3, 6.0, 4.9, 6.1, 6.2, 5.4, 4.1, 4.2, 3.8 and 3.9 tonnes. Does this fertiliser give a higher yield?

First, we should notice that in this case a one-sided test is appropriate, so we have

H_0: $\mu = 4.2$
H_1: $\mu > 4.2$

We now need the sample mean and variance

x	x^2
4.3	18.49
6.0	36.00
4.9	24.01
6.1	37.21
6.2	38.44
5.4	29.16
4.1	16.81
4.2	17.64
3.8	14.44
3.9	15.21
48.9	247.71

$$\bar{x} = \frac{\Sigma fx}{\Sigma f} = \frac{48.9}{10} = 4.89$$

$$s^2 = \frac{\Sigma fx^2}{n} - \left(\frac{\Sigma f}{n}\right)^2$$

$$= \frac{247.41}{10} - (4.89)^2 = 0.8289$$

$$s = 0.9104$$

$$\text{Test statistic } t = \frac{|4.89 - 4.2|\sqrt{(10 - 1)}}{0.9104} = 2.27$$

For a one-sided test with 9 degrees of freedom, the critical values of t are 1.833 (at the 5 per cent level) and 2.821 (at the 1 per cent level), so the sample mean is significant at the 5 per cent level but not at the 1 per cent level. We reject H_0 and accept H_1, concluding that there is evidence to suggest that the super fertiliser is effective.

When we dealt with large samples, we found it possible to test the significance of the difference between the means of two populations. Using the t-distribution, we can also do this for small samples. In some cases it is necessary to test the mean difference by calculating the difference between the items in the sample. For us to be able to do this, the samples must be 'paired', i.e. the same item must appear in both samples, though treated under different conditions. We would use this method for the 'before and after' type of statistics so favoured by advertising agencies (i.e. the samples are not independent).

Example 6

Before coaching, five candidates scored 38, 41, 52, 27 and 18 per cent respectively in a statistics examination. After coaching, they scored 40, 45, 49, 30 and 24 per cent respectively. Does this evidence suggest that coaching is effective? Although this problem appears similar to Example 3, we cannot use the same method as that method requires independent samples. In this case, we are looking at the same sample under different conditions. What should we do? Subtracting the scores, we obtain the change following coaching

$$x = 2, 4, -3, 3, 6, \qquad \bar{x} = 2.4, \qquad s = 3.007$$

Suppose we adopt as our null hypothesis that coaching is ineffective: then the mean difference could well be zero.

H_0: $\mu = 0$

We are asked to test whether coaching is effective, i.e. it increases the marks gained by the students, so a one-sided test is appropriate and the alternative hypothesis is

H_1: $\mu > 0$

Test statistic $t = \dfrac{|2.4 - 0|\sqrt{(5 - 1)}}{3.007} = 1.596$

For a one-sided test with 4 degrees of freedom, the critical values of t are 2.123 (at the 5 per cent level) and 3.747 (at the 1 per cent level). We must conclude that the test statistic is not significant, we accept the null hypothesis and reject the alternative hypothesis. There is insufficient evidence to suggest that coaching is effective.

In most cases, samples will be independent rather than paired, and if we want to test the significance of the difference between the means, we must use the method employed earlier in this chapter. Suppose we have two samples, the details of which are:

	Sample size	Mean	Variance
First sample	n_1	$\bar{x}_1$	s_1^2
Second sample	n_2	$\bar{x}_2$	s_2^2

We will assume that the first sample comes from a population with a mean μ_1, and the second sample comes from a population with a mean μ_2. We can now formulate the null hypothesis

H_0: $\mu_1 = \mu_2$ or
H_0: $\mu_1 - \mu_2 = 0$

We have a variance for each sample, and using the methods in Chapter 9 on estimation we can pool the variances of the samples to estimate the variance of the population, i.e.

$$\hat{\sigma}^2 = \frac{n_1 s_1^2 + n_2 s_2^2}{n_1 + n_2 - 2}$$

Having estimated the variance of the population, we can now estimate the standard error of the difference between two independent means to be

$$\sqrt{\frac{\hat{\sigma}^2}{n_1} + \frac{\hat{\sigma}^2}{n_2}}$$

For the two samples, the test statistic is

$$t = \frac{|x_1 - x_2| - 0}{\sqrt{\dfrac{\hat{\sigma}_2}{n_1} + \dfrac{\hat{\sigma}_2}{n_2}}}$$

Before examining an example, we must now ask ourselves how many degrees of freedom are associated with this test. There are $(n_1 - 1)$ degrees of freedom associated with the first sample, and $(n_2 - 1)$ associated with the second sample. Now as we are pooling both samples, there are $n_1 + n_2 - 2$ degrees of freedom associated with estimating the population variance.

Example 7

Lumo plc manufacture electric light bulbs, and claim that on average their lamps last longer than the lamps of their competitor, Brighto plc. A random sample of 10 lamps made by Lumo had the following lives (in hours) before failing

$x_1 =$ 200 210 190 200 190 200 180 200 200 210

and a random sample of 8 lamps produced by Brighto yielded

$x_2 =$ 190 200 210 190 180 190 200 190

Does this evidence substantiate Lumo's claim?

First, we use suitable shortcut methods to calculate the mean and standard deviation for each sample.

x_1	$\dfrac{x_1 - 200}{10}$	$\left(\dfrac{x_1 - 200}{10}\right)^2$	x_2	$\dfrac{x_2 - 200}{10}$	$\left(\dfrac{x_2 - 200}{10}\right)^2$
200	0	0	190	-1	1
210	1	1	200	0	0
190	-1	1	210	1	1
200	0	0	190	-1	1
190	-1	1	180	-2	4
200	0	0	190	-1	1
180	-2	4	200	0	0
200	0	0	190	-1	1
200	0	0		-5	9
210	1	1			
	-2	8			

$$\bar{x}_1 = 200 + \frac{-2 \times 10}{10} = 198 \text{ hours} \qquad \bar{x}_2 = 200 + \frac{-5 \times 10}{10} = 193.75 \text{ hours}$$

$$s_1^2 = 100 \times \left(\frac{8}{10} - \frac{-2^2}{10}\right) \qquad s_2^2 = 100 \times \left(\frac{9}{10} - \frac{-5^2}{10}\right)$$

$$= 76 \text{ hours} \qquad\qquad\qquad = 73.44 \text{ hours}$$

We will test the null hypothesis that there is no difference between the mean lengths of lives against the alternative hypothesis that Lumo's lamps last longer than Brighto's lamps, i.e.

$H_0: \mu_1 = \mu_2$ and $H_1: \mu_1 > \mu_2$

As in fact we are assuming that both samples could have been drawn from the same population, the estimated variance of this population is

$$\hat{\sigma}^2 = \frac{n_1 s_1^2 + n_2 s_2^2}{n_1 + n_2 - 2}$$

$$= \frac{10 \times 76 + 8 \times 73.44}{10 + 8 - 2}$$

$$= 84.22 \text{ hours, with 16 degrees of freedom}$$

The test statistic t is

$$t = \frac{|198 - 193.75| - 0}{\sqrt{\dfrac{84.22}{10} + \dfrac{84.22}{8}}} = 0.976$$

For a one-sided test with 16 degrees of freedom, the critical values of t are 1.746 (at the 5 per cent level) and 2.583 (at the 1 per cent level). So we see that the difference between the means is not significant. We accept the null hypothesis: there is insufficient evidence to suggest that Lumo's lamps last longer.

11.6   Tests of proportions

Tests involving sample proportions are extremely important in practice. In the previous chapter, we saw the basis of such a test applied to quality control. Again, many market researchers express their results in terms of proportions, e.g. '40 per cent of the population clean their teeth with Gritto'. It will be useful to design tests that will detect changes in proportions. As you can imagine, the binomial distribution forms the basis of such tests. Let us first examine how we can test the significance of the proportion of a small sample.

Example 8

Suppose it is claimed that in a very large batch of components, about 10 per cent of items contain some form of defect. It is proposed to check whether this proportion has increased, and this will be done by drawing a sample of 20 components. Adopting our null hypothesis in the usual way we have

H_0: $\pi = 0.1$
H_1: $\pi > 0.1$

Assuming that H_0 is true, we can calculate the probability distribution for samples of 20 components like this:

$$(0.9)^{20} + {}^{20}C_1(0.9)^{19}(0.1) + {}^{20}C_2(0.9)^{18}(0.1)^2 + {}^{20}C_3(0.9)^{17}(0.1)^3 + \cdots$$

Evaluating this expression we have:

No. of defectives	Probability	Cumulative probability
0	0.1215	0.1215
1	0.2700	0.3915
2	0.2850	0.6765
3	0.1900	0.8665
4	0.0897	0.9562
5	0.0319	0.9881
6	0.0089	0.9970
7	0.0020	0.9990
8	0.0004	0.9994
9	0.0001	0.9995

Examining this table, we see that the probability of obtaining more than 4 defectives is $1 - 0.9562 = 4.38$ per cent and the probability of obtaining more than 6 defectives is $1 - 0.9970 = 0.3$ per cent. So if we obtain more than 4 defectives in a sample, we would reject the null hypothesis at the 4.38 per cent level of significance and if we obtain more than 6 defectives in a sample, we would reject the null hypothesis at the 0.03 per cent level of significance.

Now calculations involving the binomial distribution are very tedious. The way to avoid them is to draw a large sample, and we can then use the normal approximation. Suppose that, in the last example, a sample of 150 contained 20 defectives – would this indicate that the proportion of defectives had increased? Again we have

H_0: $\pi = 0.1$
H_1: $\pi > 0.1$

and on the basis of our null hypothesis

mean $= n\pi = 0.1 \times 150 = 15$ defectives per sample
$\sigma \quad = \sqrt{n\pi(1 - \pi)} = \sqrt{150 \times 0.1 \times 0.9} = 3.67$

As our sample is large, our test statistic is the Z score

$$\frac{20.5 - 15}{3.67} = 1.4986$$

(Remember that we must make the 0.5 adjustment when using a normal approximation to the binomial distribution.) For a one-sided test, the critical values of Z are 1.645 (at the 5 per cent level) and 2.326 (at the 1 per cent level), so we would conclude that the proportion of defectives in the sample is not significant. We would accept the null hypothesis that the proportion defective is 0.1.

Example 9

Finally, let us examine how we could investigate the significance of the difference between proportions in two samples. Suppose that in a sample of 300 people from Liverpool, 200 cleaned their teeth with Gritto, and in Manchester 240 out of 350 use Gritto. We wish to determine whether the proportion using Gritto is the same in both cities. Clearly, this is a two-sided test, so we have

H_0: $\pi_1 = \pi_2$ (or $\pi_1 - \pi_2 = 0$)
H_1: $\pi_1 < > \pi_2$

The proportion using Gritto in Liverpool is $220/300 = 0.733$, and the proportion using Gritto in Manchester is $240/350 = 0.686$. If the null hypothesis is true, then we can regard both samples as coming from the same population. The best estimate of the proportion using Gritto in this population is

$$\pi = \frac{220 + 240}{300 + 350} = \frac{460}{650} = 0.708$$

In the chapter on sampling theory, we saw that the variance of the proportion of a sample of n items is

$$\frac{\pi(1 - \pi)}{n}$$

So the variance of the proportion of samples of 300 is

$$\frac{0.708 \times 0.292}{300}$$

and the variance of the proportion of samples of 350 is

$$\frac{0.708 \times 0.292}{350}$$

Using the theorem of sample differences, the variance of the difference between these proportions is

$$\frac{0.708 \times 0.292}{300} + \frac{0.708 \times 0.292}{350}$$
$$= 0.708 \times 0.292 \ (1/300 + 1/350)$$
$$= 0.00128$$

The standard error of the difference is $\sqrt{0.00128} = 0.0358$. Now if the null hypothesis is true, then the difference between the proportions in the samples could well be zero. The test statistic Z is

$$\frac{|0.733 - 0.686| - 0}{0.0358} = 1.31$$

For a two-sided test, the critical values of Z are 1.96 (at the 5 per cent level) and 2.576 (at the 1 per cent level), so we would accept the null hypothesis that the proportion using Gritto in Liverpool and Manchester could well be the same.

11.7 Conclusion

As you can see, there are a wide variety of significance tests for testing means and proportions. Certainly, there are many more than those mentioned in this chapter – there are tests, for example, for testing several means and proportions. However, the tests covered in this chapter will take you quite a long way in statistical analysis. Before we finish this chapter, let us again stress the importance of the sample size. You will notice that large samples impose fewer restrictions on testing, and their results will be much more reliable. It is for this reason that you should use large samples wherever possible.

Review questions

11.1 A manufacturer produces cables with a mean breaking strength of 2,000 lb and a standard deviation of 100 lb. By using a new technique, the manufacturer claims that the breaking strength can be increased. To test this claim, a sample of 50 cables produced by the new technique is tested, and the mean breaking strength found to be 2,050 lb. Can the manufacturer's claim be supported at a 1 per cent level of significance?

11.2 Consider the previous example. Could the claim be justified had the sample contained 9 cables?

11.3 A machine has been operating at 60 per cent efficiency. After adjustment, nine test runs produced the following percentage efficiencies

    64   59   71   63   68   61   62   61   63

Test, at the 5 per cent level, whether efficiency has improved.

11.4 A survey has provided data on the television watching habits of ten-year old children of different social classes. A sample of 220 children in social class five were found to watch television for a mean period of 3.8 hours per day with a standard deviation of 0.5 hours, while a sample of 63 children in social class one/two watched for a mean period of 3.2 hours with a standard deviation of 0.42 hours. Test, at the 1 per cent level, whether children in social class five spend more time watching television than do those in social class one/two.

11.5 In a certain country, men have a mean height of 5 ft 6 in, standard deviation 3 in, and women have a mean height of 5 ft 3 in, standard deviation 2 in. In a random sample of 100 married couples, the average height difference is 2 in. Does this suggest that the height of a partner affects the decision to propose marriage?

11.6 In order to test the efficiency of a drying agent in paint, the following experiment was carried out. Each of six samples was cut in two halves. One half was covered in paint containing the drying agent and the other half with paint without the agent. All twelve samples were left to dry. The time taken to dry was as follows:

Drying time (hours)	1	2	3	4	5	6
Paint with agent	3.4	3.8	4.2	4.1	3.5	4.7
Paint without agent	3.6	3.8	4.3	4.3	3.6	4.6

Carry out a t-test to determine whether the drying agent is effective. Give your reason for choosing a one-sided or two-sided test. Carefully explain your conclusions.

11.7 The following data refer to the number of completed dwellings over a period of 11 years.

Year	Private dwellings (thousands)	Local authority dwellings (thousands)
1	177.8	123.9
2	221.3	154.7
3	217.2	165.0
4	208.6	176.9
5	204.2	199.7
6	226.0	188.0
7	186.0	180.9
8	174.3	177.0
9	196.3	154.8
10	200.6	120.4
11	190.6	102.6

Is there a significant difference in the average number of dwellings completed in the two groups over the 11 years?

11.8 Suppose samples of 10 items are to be drawn from a population with 5 per cent defective items. Find the approximate 5 per cent and 1 per cent critical values for the alternative hypothesis that the proportion defective has increased.

11.9 A company selling a prestige product carries out market research surveys in two consecutive years on heads of households with incomes in excess of £10,000 per year. The survey results were:

	Year 1	Year 2
Sample size	1,300	1,000
Number possessing product	351	240

Using a 5 per cent level of significance, does the year 2 survey suggest that the product's sales are declining?

11.10 Two groups, A and B, each consist of 100 people who have a certain disease. A serum is given to group A but not to group B: otherwise the two groups are identical. It is found that in groups A and B, 75 and 65 people respectively recover from the disease. Does this result support the hypothesis that the serum helps cure the disease?

12 The chi-square test

12.1 Introduction

We will now pick up and develop a point made in the previous chapter – that in all the tests we have studied so far, we have assumed that the population from which the sample was drawn was normally distributed and that the sample we drew was truly random. Even if the population was not normally distributed, we learnt from the central limit theorem that the sampling distribution of sample means would be normally distributed. Thus, whether we are considering a population known to be normally distributed or are considering a sample drawn from a population with unknown characteristics, we have based our tests and estimates on the normal distribution. Even when we discussed small samples, and used the t-test, we were making an implicit assumption that, provided our population does not depart too far from the normal distribution, the t-test will yield valuable results. All the empirical evidence we have tends to support the view that the t-test is insensitive to movements away from the normal distribution when we are testing null hypotheses of sample means.

Now, why do we make this assumption of a normally distributed population when in fact such populations are very rare? The point is, all the tests we have looked at involve the use of parameters such as the population mean, the population proportion and the population variance. If we assume that the population is normally distributed we can infer a great deal from these parameters. You will remember, for example, that the variance of the sampling distribution of sample means (the standard error) is taken to be equal to $\sigma^2/\sqrt{n}$, or that 95 per cent of the normally distributed population will fall within the range $\mu \pm 1.96\sigma$.

Suppose, however, that we do not know, and are unwilling to assume, the precise form of the population distribution. We can not now make any assumptions about the parameter values. Yet often a statistical test would appear to be useful. In such circumstances we would have to use a *distribution free*, or *non-parametric*, test technique. You will understand, of course, that even if we could confidently make presumptions about the population distribution or parameters, we need not do so.

The methodology we use for non-parametric tests is identical to that which we have already used for both Z and t tests. First, on the basis of a null hypothesis, a sampling distribution of some sample characteristic is derived. Then, having prescribed an acceptable level for a Type 1 error a decision criterion can be fixed. If the sample characteristic exceeds the critical value it is deemed significant at the level you have chosen; if it does not fall into the region of rejection the null hypothesis is accepted, subject always to the possibility that a Type 2 error may have been committed. And this is the point. For samples of the same size, non-parametric tests always leave a higher possibility of a Type 2 error than do parametric tests. We can always offset this, of course, by taking a larger sample size. In any case non-parametric tests make up in generality what they lack in power. The conclusions we reach do not depend on the assumptions which may only apply to the situation we are studying, and so they tend to be of universal application.

12.2   The chi-square test

Probably the most widely used of all non-parametric tests is the chi-square test (pronounce it 'kye' and use the symbol χ^2), and certainly it has a wider application than most other tests. Like the t-distribution, there are many χ^2 distributions – one for each degree of freedom (v). The fewer the degrees of freedom, the more positively skewed is the distribution. Conversely, the greater the number of degrees of freedom the closer does the χ^2 approximate to the normal distribution.

12.3   Observed and expected frequencies

Suppose you have been given the task of assessing consumer preference with respect to five brands of washing powder – Zip, Whito, Acme, Blanco and Bleacho. You supply a quantity of each powder in boxes marked A, B, C, D and E to 1,000 housewives, ask them to test the powders and state which powder they prefer. The results of your enquiry is as follows:

Washing powder	Zip	Whito	Acme	Blanco	Bleacho
Number preferring that powder	187	221	193	204	195

Now you suspect that housewives are unable to distinguish between washing powders. Any preference that they have will be influenced not by the powder's characteristics, but by the persuasive powers of the advertisements promoting the powder, and the attractiveness of the packet. For this reason, the housewives do not know which powder they are using (remember they are labelled A, B, C, D and E, and that each powder is packed in a plain white box). Returning to statistical jargon, we have formulated the null hypothesis that housewives are unable to distinguish between the powders, and we will use sample evidence to test the validity of this hypothesis. Of course, the appropriate alternative hypothesis is that the housewives can distinguish between the powders.

Earlier in this section, we implied that the null hypothesis tested should always be formulated in such a fashion that we would know what to expect if it was true. Assuming that our null hypothesis is correct, then, we expect that an equal number would report that they preferred each of the powders. In other words, the results expected from our sample would be:

Washing powder	Zip	Whito	Acme	Blanco	Bleacho
Number expected to prefer that powder	200	200	200	200	200

The frequencies that we obtain in our sample we will call the *observed frequencies*, and the frequencies we expect on the basis of our null hypothesis we will call the *expected frequencies*. We shall now devise a significance test on whether there is a significant difference between the observed and expected frequencies.

The method we shall apply to testing the null hypothesis is to measure the total deviation of the observed from the expected frequencies, i.e. $\Sigma\,(O - E)$.

Product	Observed	Expected	$(O - E)$
Zip	187	200	-13
Whito	221	200	21
Acme	193	200	-7
Blanco	204	200	4
Bleacho	195	200	-5
			0

As you would expect $\Sigma\,(O - E) = 0$, so we are faced with a problem. If $\Sigma\,(O - E)$ always equals zero, then we cannot use it as a measure for testing our null hypothesis. The way out is to square the values $(O - E)$, as a negative number squared becomes positive and so the total deviation squared will not be zero. However, a further adjustment is necessary: we divide the $(O - E)^2$ values by E. Now why do we do this? Well, the reason is not apparent in this example, so let us consider another case.

Observed	Expected	$(O - E)$	$(O - E)^2$
150	200	-50	2,500
150	100	50	2,500

Clearly, the deviation of 50 is twice as significant on an expected value of 100 than it is on an expected value of 200, and to take this into account, we can weight the $(O - E)^2$ values by dividing by E. So we have

Observed	Expected	$(O - E)$	$(O - E)^2$	$\dfrac{(O - E)^2}{E}$
150	200	-50	2,500	12.5
150	100	50	2,500	25

Which shows precisely the information required. Now if we calculate the sum of $(O - E)^2/E$ values, we have a weighted sum of the deviations of the observed from the expected values, and we call this statistic χ^2.

$$\chi^2 = \Sigma\,\frac{(O - E)^2}{E}$$

Calculating the value for χ^2 for our washing powder example we have

Observed	Expected	$(O - E)$	$(O - E)^2$	$\dfrac{(O - E)^2}{E}$
187	200	-13	169	0.845
221	200	21	441	2.205
193	200	-7	49	0.245
204	200	4	16	0.080
195	200	-5	25	0.125
				3.500

If our null hypothesis is true, then our expected value for χ^2 is zero, as we would expect our observed and expected values to have the same value. But as we are sampling, the actual values of χ^2 will fluctuate from sample to sample. Now χ^2 is like Z in so far as it is just another kind of standard score — so to test our null hypothesis we can use the same method as we did in the previous chapter. We will compare our value for χ^2 with certain critical values that we would not expect to be exceeded. However, there are two problems involved in obtaining these critical values.

When we examined sampling distributions, we learnt that as long as the sample size was sufficiently large, then we can expect the sample statistic to be normally distributed. We could use the normal tables to obtain critical values, and compare them with the Z score of the sample statistic. Now here is where the first problem occurs – χ^2 is not normally distributed, so the normal distribution tables are of no help to us. We must obtain our critical values from χ^2 tables. Now there are many χ^2 distributions, but only one standard normal distribution, so here is where our second problem arises – which χ^2 distribution should we use?

Just as with the t-distribution there are many χ^2 distributions – one for each degree of freedom. The distribution applicable in this case is the one with 4 degrees of freedom – can you see why? Clearly, when we assigned expected values to the numbers buying each detergent, we were not completely free to assign any values that we wished. In order that we can compare expected values with observed values, the total of the expected values had to be 1,000. This means that any hypothesis we adopt leaves us free to assign expected values to any four of the five brands. For example, we could have assigned an expected frequency of 150 to Whito, 300 to Acme, 200 to Blanco and 150 to Bleacho – then the expected value for Zip *must* be $1,000 - (150 + 300 + 200 + 200 + 150) = 300$, otherwise we would not have a total frequency of 1,000. So we have just 4 degrees of freedom when fixing expected values. (Generalising for problems like this, if we have n observed values, then we will have $(n - 1)$ degrees of freedom in calculating the expected values.)

Consulting the χ^2 tables with 4 degrees of freedom, then we see that the critical values are 9.49 (at the 5 per cent level) and 13.3 (at the 1 per cent level). As our value of χ^2 is 3.3 (much less than both critical values), we would accept the hypothesis that housewives are unable to distinguish between the washing powders.

12.4 Errors resulting from the use of χ^2

Before we go any further, it is worth considering the errors that can occur if we use χ^2. One possible source of error is that the χ^2 distribution is continuous, though we have used it to test data that is discrete. Now this error is not really serious *unless there is only 1 degree of freedom*, and when $v = 1$ we compensate for this error by applying *Yate's correction*. All that is involved in this correction is that we subtract 0.5 from the *absolute* difference between the observed and expected values. Let us examine an example to see how we apply the correction.

Example 1

In a sample of 1,000 people, 452 said they would vote Socialist at the next election. We wish to test the hypothesis that 50 per cent of the electorate vote Socialist. On the basis of this hypothesis, then, we have:

	Observed	Expected
Socialist	452	500
Not Socialist	548	500

In this case we have only 1 degree of freedom, so our critical values are 3.84 (at the 5 per cent level) and 6.63 (at the 1 per cent level). Also, as $v = 1$ we must apply Yate's correction.

Observed	Expected	$(O - E)$	$(\lvert O - E \rvert - 0.5)$	$(\lvert O - E \rvert - 0.5)^2$	$\dfrac{(\lvert O - E \rvert - 0.5)^2}{E}$
452	500	-48	47.5	2,256.25	4.5125
548	500	48	47.5	2,256.25	4.5125
					$\chi^2 = 9.0250$

So we can see that the sample result is significant at the 1 per cent level. We reject the null hypothesis and conclude that less than 50 per cent of the electorate vote Socialist.

There is a second type of error that can arise from using the χ^2-distribution. If the expected frequency of any cell is unusually small, then the use of χ^2 may lead to erroneous conclusions. Two general rules for small-cell frequencies are:

(a) If there are only two cells, the expected frequency of each cell should be five or more. We could then use the χ^2 test in the following case:

	Observed	Expected
Agreeing	643	642
Disagreeing	4	5

but not in the following case:

	Observed	Expected
Agreeing	643	644
Disagreeing	4	3

(b) If there are more than two cells, χ^2 should not be used if more than 20 per cent of the expected frequencies are less than five. Consult the table below. According to this rule we could use the χ^2 test for the examination data on the left. Only one cell out of six has a frequency less than five, i.e. only 16.67 per cent of cells. The χ^2 test should not, however be used for the data on the right, since three of the seven cells have an expected frequency of less than five, i.e. 43 per cent.

No. of students		Examination	No. of students	
Observed	Expected	Grade	Observed	Expected
18	16	A	30	32
39	37	B	110	113
8	13	C	86	87
6	4	D	23	24
82	78	E	5	2
10	15	F	5	4
–	–	U	4	1
163	163		263	263

The reasoning behind the second rule can be seen if we examine the data carefully. You will see that there is a close relationship between the observed and expected frequencies. The maximum difference is three. It would be sensible to assume that there is no significant difference between the observed and expected frequencies. Yet if you compute the value of χ^2 you will find it to be 14.01, greater than the critical value at the 5 per cent level. Contrary to common sense, we would reject the null hypothesis. The dilemma can be resolved if we combine the last three classes into a single class:

Grade	Observed	Expected
E and below	14	7

Using these revised frequencies the computed value of χ^2 is 7.26, less than the critical value for

4 degrees of freedom at the 5 per cent level (9.49). The null hypothesis would be accepted, and we would argue that there is no difference between the observed and expected frequencies. This is, of course, a much more logical conclusion.

12.5 Contingency tables

To illustrate a further application of χ^2 testing, let us now examine a question set by ACCA.

Example 2

Using the χ^2 distribution as a test of significance, test the statement that the number of defective items produced by two machines, as shown in the following table, is independent of the machine on which they were made.

| | Machine Output | | |
	Defective articles	Effective articles	Total
Machine A	25	375	400
Machine B	42	558	600
	67	993	1,000

We are asked to test the statement that the number of defective items is independent of the machine on which they were made, and we adopt this as our null hypothesis. Now if this statement is not true, then the number of defectives will depend upon the machine on which they were made, and the table will enable us to calculate the degree of dependence. A table constructed in this way (to indicate dependence or association) is called a *contingency table*. 'Contingency' means dependence – many of you, for example, will be familiar with the term 'contingency planning', i.e. plans that will be put into operation *if* certain things happen.

As usual, let us suppose that our null hypothesis is true. Given this assumption, then we are only interested in the totals in the table, i.e. that out of 1,000 articles, 67 were defective, and that out of 1,000 articles, 400 were produced on machine A. Using this information, we can now predict the expected number of defectives produced by the two machines. We have 400/1,000 = 4/10 of the total output produced on machine A, and as 67 items are defective, we would expect $67 \times 4/10 = 26.8$ of them to have been produced on machine A. Now as machine A produces 400 items, then $400 - 26.8 = 373.2$ items can be expected to be non-defective. Likewise, if 26.8 of the 67 defectives are produced on machine A, it follows that $67 - 26.8 = 40.2$ of them will have been produced on machine B. Finally, as we can expect machine B to produce 600 items of which 40.2 will be defective, then $600 - 40.2 = 559.8$ can be expected to be non-defective. The expected table would look like this:

| | Machine Output | | |
	Defective articles	Effective articles	Total
Machine A	26.8	373.2	400
Machine B	40.2	559.8	600
	67.0	993.0	1,000

Notice that we have calculated just one of the values in the table, the remainder will be fixed by subtraction. So we have $v = 1$ degree of freedom in calculating the expected values for a 2×2

contingency table and we must apply Yate's correction.

Observed	Expected	$(O - E)$	$(\|O - E\| - 0.5)$	$\dfrac{(\|O - E\| - 0.5)^2}{E}$
25	26.8	-1.8	1.3	0.063
375	373.2	1.8	1.3	0.005
42	40.2	1.8	1.3	0.042
558	559.8	-1.8	1.3	0.003
				$\chi^2 = 0.133$

With $v = 1$ we would reject the null hypothesis if $\chi^2 > 6.64$ (at the 1 per cent level). Now as our value is much less than this, there is no reason to reject the null hypothesis – the number of defective items is independent of the machine on which they were produced.

If we are going to use the above technique, then we must make absolutely sure that the samples are *independent*. We could not use it if the same sample is being tested under different conditions (i.e. 'before and after' investigations). The following example will illustrate this.

Example 3

A sample of 100 housewives were given a standard washing powder to test and report whether the powder was satisfactory or unsatisfactory. The same women were then asked to test a new super powder. The results were:

		New powder	
		Satisfactory	Unsatisfactory
	Satisfactory	20	10
Old powder			
	Unsatisfactory	50	20

Examining this table, we see that 60 women 'change their minds', i.e. 50 changed from unsatisfactory to satisfactory, and 10 changed from satisfactory to unsatisfactory. Now suppose we formulate the null hypothesis that the type of powder had no effect on their response (i.e. they could not distinguish between them), then we would expect an equal number to change their minds in either direction. So we would expect 30 to change from satisfactory to unsatisfactory and 30 to change from unsatisfactory to satisfactory. This now enables us to calculate χ^2.

Observed	Expected	$(O - E)$	$(\|O - E\| - 0.5)$	$\dfrac{(\|O - E\| - 0.5)^2}{E}$
50	30	20	19.5	12.675
10	30	-20	19.5	12.675
				$\chi^2 = 25.350$

Using the critical values for $v = 1$, we must reject the null hypothesis at the 1 per cent level. The difference between the observed and expected values cannot be attributed to chance and we must conclude that association between response and type of powder is established.

Finally, let us see how we can test for association with larger than 2×2 tables.

Example 4

Suppose that a random sample of men and women indicated their view on a certain proposal as follows:

	In-favour	Opposed	Undecided	
Women	118	62	25	205
Men	84	78	37	199
	202	140	62	404

Test the hypothesis that there is no difference in opinion between men and women as far as this proposal is concerned.

We adopt the null hypothesis that there is no association between the response and the sex of the person interviewed. On this basis we many deduce that the proportion in the sample who are female is 205/404, and as 202 people are in favour of the proposal, the expected number of women in favour of the proposal is

$$205/404 \times 202 = 102.5$$

Also, as 140 people are against the proposal, the expected number of women against the proposal is

$$205/404 \times 140 = 71$$

We can now obtain the remaining values by subtraction, i.e.

Expected number of undecided women $= 205 - (102.5 + 71) = 31.5$
Expected number of men in favour $= 202 - 102.5 = 99.5$
Expected number of men opposed $= 140 - 71 = 69$
Expected number of undecided men $= 62 - 31.5 = 30.5$

So the table would look like this:

	In-favour	Opposed	Undecided
Women	102.5	71	31.5
Men	99.5	69	30.5

We now have all the necessary information to calculate the value of χ^2:

Observed	Expected	$(O - E)$	$\dfrac{(O - E)^2}{E}$
118	102.5	15.5	2.34
62	71	-9	1.14
25	31.5	-6.5	1.34
84	99.5	-15.5	2.41
78	69	9	1.17
37	30.5	6.5	1.38
			$\chi^2 = 9.78$

In this example, we have 2 degrees of freedom in calculating the expected values. Consulting the table, we see that the critical values for χ^2 are 5.99 (at the 5 per cent level) and 9.21 (at the 1 per cent level), hence we would reject the null hypothesis accepting the alternative hypothesis (that men and women think differently) with 99 per cent confidence.

12.6 Conclusion

We can see, then, that the χ^2 test is very versatile, and it might be useful to summarise how to find the number of degrees of freedom.

  (a) For goodness of fit tests with n classes, $v = n - 1$
  (b) For tests of association with m rows and n columns $v = (m - 1)(n - 1)$

It should be noted that of all significance tests, χ^2 is probably the most widely used. However, many statisticians are highly critical of χ^2 testing, reasoning that there are other tests available which are far more suitable. Although this is not the place to discuss such criticisms in detail, there is one point that you may consider. None of the null hypotheses we have considered with respect to goodness of fit can be *exactly* true, so if we increase the sample size (and hence the value of χ^2) we would ultimately reach the point when all null hypotheses would be rejected. All the χ^2 test can tell us, then, is that the sample is too small to reject the null hypothesis!

Review questions

12.1 The number of rejects in six batches of equal size were

Batch	Number of rejects
A	270
B	308
C	290
D	312
E	300
F	320

Test the hypothesis that the differences between them are due to chance. Use a 5 per cent level of significance.

12.2 The number of breakdowns that have occurred during the last 100 shifts is as follows.

Number of breakdowns per shift	0	1	2	3	4	5
Frequency						
Expected number of shifts	14	27	27	18	9	5
Actual number of shifts	10	23	25	22	10	10

Show whether the manager is justified in his claim that the difference between the actual and expected number of breakdowns is due to chance. Use a 5 per cent level of significance.

12.3 A manufacturer of fashion garments for the younger age groups suspects that the market for his product has changed recently. Sales records for previous years showed that 14 per cent of buyers were below 16 years of age, 38 per cent were 16 to 20 years of age, 26 per cent were 21 to 25 years of age, and 22 per cent were over 25. A random sample of 200 recent buyers, however, showed the following results:

Age	Under 16	16–20	21–25	above 25
Frequency	22	62	60	56

Required:

  (a) Compare the results of the sample with those expected from previous records.
  (b) What null hypothesis should be tested using this data?
  (c) Carry out a chi-square test of significance.
  (d) Express your conclusions in terms meaningful to management.

12.4 A market research agency has been commissioned to report on the smoking habits of men and women in a particular area. Two samples were taken, one of 200 men and the other of 200 women. The findings are shown in the following table:

	Smokers	Non-smokers	Total
Men	110	90	200
Women	104	96	200

You are required to:

(a) test the hypothesis that difference in sex has no effect on the number of smokers in the two samples;

(b) explain your conclusions in (a).

12.5 The following data is based on a random sample of 1,000 businessmen.

	Profit maximisers	Revenue maximisers	Total
Studied economics	200	100	300
Did not study economics	200	500	700

Test whether the above information suggests any association between education in economics and the objectives of businessmen.

12.6 A certain drug was administered to 100 volunteers with common colds, and 35 reported that the drug brought some relief. The next winter, a new improved drug was given to the same group, and this time 43 reported that the drug brought some relief. Are we justified in assuming that the second drug is more likely to bring relief?

12.7 Pharmaceuticals plc have developed three new respiratory drugs. They submitted the drugs to clinical trials, the results of which are given in the table below. Determine whether or not these figures indicate a real difference in response of the disease to the drugs, and explain your conclusions. (Use a 1 per cent level of significance.)

Drug	Percentage of patients recovered	Percentage of patients not recovered
A	65	35
B	90	10
C	85	15

12.8 Two factories, using materials purchased from the same supplier and closely controlled to an agreed specification, produce output for a given period, classified into three quality grades as follows:

	Output in tons			
Quality grade:	A	B	C	Total
Factory:				
X	42	13	33	88
Y	20	8	25	53

Do these output figures show a significant difference at the 5 per cent level?

Part Four   Analytical techniques

13   Inventories

13.1   Introduction

No firm can operate without holding stocks, even if stocks held are small, or are they held for a short period of time. Manufacturers need to hold stocks of raw materials or semi-finished goods in order to maintain a smooth flow of production. They also need to hold stocks of finished goods to meet the requirements of their customers. Moreover, over the last fifty years the the trends towards specialisation have intensified. Consider, for example, the motor car manufacturer. Apart from the actual shell of the car, his contribution to production is merely to assemble together those components produced by other manufacturers. Clearly, this trend to specialisation has increased the importance of a rational policy towards stock – or, as this is sometimes called, an *optimum inventories policy*. The sort of problems we shall consider in this chapter will be the quantity of stock to be ordered, and the time interval between orders. To decide on an optimum inventories policy, however, we need some criterion to enable us to decide that the policy chosen is, in fact, the best one. Surely, the best policy is the one that involves the lowest inventory costs?

13.2   Inventory costs

It seems appropriate at this point to attempt classification of inventory costs. First, we shall consider *holding costs* which are those which arise directly from a decision to hold stocks. How do such costs arise? Well, if stocks are held, it will be necessary to keep them somewhere – perhaps in a warehouse. Thus, holding stocks may involve building or buying premises, and heating and lighting for such premises. It involves paying warehouse staff to keep track of the stock, and to move it to where it is needed, when it is needed. Somehow, all such costs must be apportioned to each unit of stock. Fortunately, this is an exercise in cost accountancy, and need not concern us here.

In addition to warehouse costs, there is a second and highly important element in holding costs. Buying inventories ties up a firm's capital – and this capital could be used to earn interest. Suppose a firm decides to buy £1,000 worth of spare parts for its machinery. The spare parts are then put into stock until needed. Now instead, the firm could have invested this £1,000 and earned interest. At 10 per cent per annum, this £1,000 would have earned £100 interest in a year. It is reasonable, then, to think of this £100 that the firm could have earned as being an inventory cost – it is money that the firm has forgone through its decision to hold stocks.

Summarising, holding costs comprise storage cost plus cost of capital tied up. Holding costs are expressed as a value 'per unit per time period'. Suppose, for example, it costs 10p to store a certain component per month. The component cost £10, and the cost of capital is 15 per cent per year. It costs 10p × 12 to store the component for a year, and additionally the firm has sacrificed 15 per cent of £10 by tying up capital. Thus, the holding cost is $0.1 \times 12 + 10 \times 0.15 = £2.70$ per unit per year.

A second type of inventory cost is *ordering costs* – costs that are incurred each time an order is placed. By far the most obvious element in ordering cost is the clerical cost involved in actually preparing and making the order. There is also the task of ensuring that the best price is being obtained. Ordering costs are expressed as so much 'per order'.

The final element in inventory costs is *stockout costs*, i.e. those which are incurred when firms actually run out of stock. Sometimes, stockout costs are easily calculated: it may be the cost of machines and men being idle through lack of components or raw materials or spare parts; it may be the loss of profit through a failure to fulfil an order. However, running out of stock can cause loss of goodwill – certainly a cost, but very difficult to evaluate!

13.3   An inventory model

Tiger Sports Cars produce a high-performance, rather expensive sports car. Now most of the components for this car are not produced by Tiger itself but, in common with the volume producers, are bought from specialist producers. Obviously, Tiger must have stocks of components, and it will be most interesting to examine the factors that determine the size of such stocks. In order to do this, we shall make three assumptions. First, we shall assume that production occurs at a steady rate throughout the year; in other words, we are assuming linear demand for components. Secondly, we will assume that any breaks in production are so expensive that the firm would move heaven and earth to ensure that it never runs out of components. Using the mathematician's jargon, we are assuming that stockout costs are infinitely high, and later we will examine the implications of this assumption. Finally, we shall assume that there is no time lag between placing an order for components and receiving delivery (this reinforces the second assumption). This rather artificial assumption enables the firm to reorder when stock runs out, but still avoid stockouts! Armed with these assumptions, we can represent the levels held by the firm by Fig.13.1.

Here, the firm is placing regular orders for q items with its supplier. The straight line AB shows that stock is used up at a steady rate for a time period t, after which the stock is exhausted (this time period t is called the inventory cycle). Stock is then replenished immediately. The problem we shall be considering is the optimum level of q, i.e. the level of q that minimises inventory costs (the optimum level of q is frequently called the *economic ordering quantity*, so you will often find inventory problems referred to as EOQ models).

To find the optimum batch size to order, we need more information. We will need to know how many components are required per time period, and we will need to know the inventory costs. Let

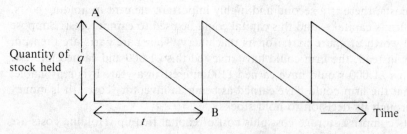

Figure 13.1

us suppose Tiger produces 5,000 vehicles per year and that we are considering an ordering policy for one component — headlights. Clearly, Tiger will require 10,000 headlamps per year. If the firm decides to order in batches of 1,000 headlamps, then $10,000 \div 1,000 = 10$ orders per year must be placed. Again, if the firm orders in batches of 100, then $10,000 \div 100 = 100$ orders per year must be placed. Generalising, if the firm orders batches of q items, then:

$$\frac{10,000}{q} \text{ orders per year will be placed}$$

The annual ordering cost can be calculated by multiplying the cost of one order by the number of orders placed per year. Tiger's accountants have calculated that it costs £8 to place an order for headlamps, so:

$$\text{Ordering costs} = \frac{10,000}{q} \times 8 = \frac{£80,000}{q}$$

Figure 13.2 shows clearly the relationship between the batch size ordered and annual ordering costs. In the table below, arbitrary values for the batch size q have been chosen, and the associated annual ordering costs calculated:

if batch size (q) =	50	100	150	200	250	300	350	400
then ordering costs per annum (£) =	1,600	800	$533\frac{1}{3}$	400	320	$266\frac{2}{3}$	228.6	200

A glance at Fig. 13.2 shows that as we increase the quantity ordered, so we will reduce the annual ordering costs. Now this is precisely what we would expect – larger orders means fewer orders, and fewer orders means lower annual ordering costs.

So much for ordering costs: we shall now turn our attention to holding costs. The accountants have calculated that it costs £2 per year to store a headlamp. If headlamps cost £10 each, and the cost of capital is 20 per cent per annum, then it costs $£10 \times 0.2 = £2$ in tied-up capital to hold a headlamp for a year. So the total, annual holding cost per year for one headlamp is £2 (storage) + £2 (in tied-up capital) = £4. As ordering costs are expressed in terms of 'total per year', then it would be advisable

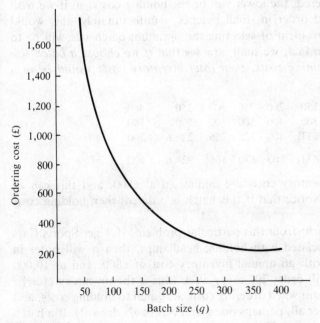

Figure 13.2

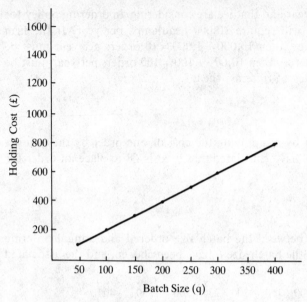

Figure 13.3

to do the same with holding costs. If the firm regularly orders in batches of q items, then the average stock held would be $q/2$, and the total, annual holding cost would be $q/2 \times £4 = £2q$.

Again, it would be illuminating to draw a diagram, this time showing the relationship between the batch size ordered and the annual holding costs (Fig.13.3):

if the batch size (q) = 50   100   150   200   250   300   350   400
then holding costs (£) = 100   200   300   400   500   600   700   800

Notice that the smaller the batch size ordered, the lower will be the holding costs, so if we wish to reduce the holding costs, then we would order in small batches – unfortunately, this would increase ordering costs! Clearly, then, the problem of selecting the optimum batch size will be to balance ordering costs with holding costs. In fact, we shall now see that *if we choose a batch size that exactly balances ordering costs and holding costs, then total inventory costs would be at a minimum*.

Batch size (q)	50	100	150	200	250	300	350	400
Annual holding cost (£)	100	200	300	400	500	600	700	800
Annual ordering cost (£)	1,600	800	$533\frac{1}{3}$	400	320	$266\frac{2}{3}$	228.6	200
Total cost	1,700	1,000	$833\frac{1}{3}$	800	820	$866\frac{2}{3}$	928.6	1,000

We can see from Fig. 13.4 that total inventory costs are minimised at £800, and this can be achieved by ordering batches of 200 items. Notice that if this batch is ordered, then holding costs and ordering costs are both £400.

We shall now summarise what we have learnt from this particular problem. If Tiger Sports Cars wishes to minimise the inventory costs associated with holding headlamps, then it will order in batches of 200. This will involve the firm with an annual inventory cost of £800, and as 10,000 headlamps per year are required, the firm will order 200 every week (this is the inventory cycle).

We saw that inventory costs are at a minimum when ordering costs are equal to holding costs, and it would be as well be check that this is so (after all, perhaps our graph was badly drawn!). If a batch size of 200 gives a minimum level of inventory costs (£800), then any departure from 200 will involve

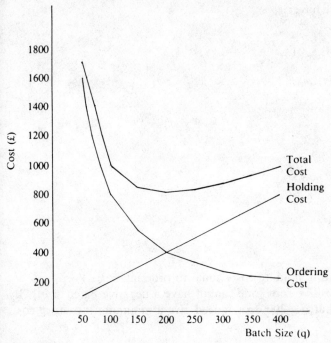

Figure 13.4

costs greater than £800. Batches of 201 would involve annual inventory costs of

$$\frac{80,000}{201} + 2 \times 201 = £800.00995$$

(ordering) + (holding) = (total)

and batches of 199 would involve annual inventory costs of:

$$\frac{80,000}{199} + 2 \times 199 = £800.01005$$

So we can see that even marginal changes from our optimum batch size will involve marginal increases in total annual inventory costs.

We have now learnt something extremely useful: to find the optimum batch size for problems like the one facing Tiger, there is no need to draw the appropriate graphs – simply put ordering costs equal to holding costs and solve for q. In this case, we have:

$$\frac{80,000}{q} = 2q$$

$$2q^2 = 80,000$$

$$q^2 = 40,000$$

$$q = \pm 200$$

Clearly, it is the positive root that makes sense in this particular situation.

If you remember the principles of differentiation, then you might feel (with some justification) that we have made a meal of solving this problem. There is no need to graph the functions: all that we need to do is to differentiate the cost function with respect to the batch size, and put the derivative equal to zero. Calling C the total cost, we have:

$$C = \frac{80,000}{q} + 2q$$

$$\frac{dC}{dq} = \frac{-80,000}{q^2} + 2$$

For a minimum value of C,

$$\frac{dC}{dq} = 0, \text{ i.e.}$$

$$\frac{-80,000}{q^2} + 2 = 0$$

$$2q^2 = 80,000$$

$$q = \pm 200$$

(Notice this is exactly what we got by putting holding costs equal to ordering costs.)

Now clearly, we are interested in the positive root (one cannot have a negative batch size!). By taking second derivatives, we can check that $q = 200$ does indeed give a minimum inventory cost:

$$\frac{d_2 y}{dq^2} = \frac{160,000}{q^3}$$

and with $q = 200$, the second derivative will have a positive value, i.e:

$$\frac{160,000}{(200)^3} > 0$$

so $q = 200$ does give a minimum inventory cost.

Which method should you use? Well, it is really up to you. We feel if you have mastered an elementary knowledge of calculus, then you would be well advised to solve inventory problems using differentiation – after all, it is *the* method of solving problems of maxima or minima. Also, if you ever meet more complex inventory problems, then you will *have* to use calculus to solve them. If the process of differentiation is a closed book as far as you are concerned, then use the method applied earlier, i.e. put ordering costs equal to holding costs and solve for q. But do try to appreciate that by using this method, your appreciation of inventory problems will be somewhat limited.

As an alternative to both working from first principles or to differentiating, you can instead use a formula to determine the optimum batch size. Suppose we let q = the optimum batch size, d = the units demanded per time period, cs = the cost of placing an order, and cl = the cost of holding a unit of inventory per time period, then:

$$q = \sqrt{\frac{2 \times cs \times d}{cl}}$$

so in our example, $d = 10,000$ per year, $cs = £8$ per order and $cl = £4$ per year:

$$q = \sqrt{\frac{2 \times 8 \times 10,000}{4}}$$

$$= 200 \text{ (as before)}$$

The number of orders placed would be $10,000/200 = 50$ (i.e roughly one per week). The ordering cost would be $50 \times £8 = £400$ and the holding cost would be $200/2 \times 4 = £400$. So the total annual inventory cost would be £800 per year. As was stated before, whichever method you use is up to you, and you will obviously choose the one that comes easiest to you. However, before you try the review questions, a word of warning: read the questions carefully and make sure that the ordering cost and the holding cost both refer to the same time period. Also, you should ensure that all costs are in the same monetary units (i.e. all pounds or all pence).

13.4   The effect of discounts

We have seen that it is logical for Tiger to order in batches of 200 headlamps, because this minimises inventory costs. However, the supplier of headlamps may not consider this to be a convenient batch size, and may well experience economy of handling if only he or she could supply headlamps in larger batches. Regular orders of (say) 1,000 headlamps might make it worthwhile for the supplier to install mechanised handling facilities (for example, the use of fork-lift trucks). Now one way that the supplier might induce Tiger to order larger batches is to offer a discount for larger orders. Suppose Tiger is offered a 5 per cent discount for orders of 1,000 headlamps – should Tiger accept this offer?

Let us first recap on order levels of 200 components. This will involve Tiger with inventory costs of £800. In addition to this, however, Tiger must spend $10,000 \times £10 = £100,000$ on actually buying the headlamps, so the total outlay on headlamps would be £100,000 + £800 = £100,800 per year. Clearly, if Tiger is to be induced to accept the supplier's offer, then it must involve Tiger in an annual outlay of less than £100,800 per year.

Suppose Tiger did order batches of 1,000 items. Ordering costs would still be £80,000/q, and because the batch size ordered has increased to 1,000, ordering costs would fall to 80,000/1,000 = £80 (previously, they were £400). Holding costs would no longer be 2q because the cost of capital tied up will have fallen (headlamps will now cost £9.50 rather than £10). The holding cost per unit per year would be £2 (storage cost) + 0.2 × £9.50 (cost of capital) = £3.90. So the total holding cost per year would be $q/2 \times 3.90 = 1.95q$, and with batches of 1,000, the annual holding cost would rise to $1,000 \times 1.95 = £1,950$ (previously it was £400). So the total inventory cost per year would be £1,950 + £80 = £2,030. So if batches of 1,000 were ordered rather than batches of 200, then total inventory cost would rise by 2,030 − 800 = £1,230. But we must not forget that by ordering batches of 1,000, Tiger is entitled to a 5 per cent discount – a saving of 50p per headlamp. So the total saving made on the 10,000 headlamps bought in a year would be £5,000. Obviously, Tiger should be advised to take advantage of the discount offered and order batches of 1,000 headlamps.

Tiger makes substantial savings – £3,770 per year – by increasing the size of batches ordered to 1,000 in return for a 5 per cent discount. It would appear that Tiger would have been willing to take up this offer if the discount was less than 5 per cent, and it would be interesting to know what would be the minimum discount that Tiger is prepared to consider. To do this, let us call the discounted price p.

Ordering cost would be £80,000/1,000 = £80. Storage cost remains at £2. Cost of capital tied up is 20 per cent of the discounted price, i.e. 0.2p. Average stock held is 1,000/2 = 500. So annual holding cost is:

$$500 \times (2 + 0.2p) = 1,000 + 100p$$

The total cost of purchasing headlamps is 10,000p. The total outlay on inventory is:

$$80 + 1,000 + 100p + 10,000p$$
$$= 1,080 + 10,100p$$

If Tiger is to take up the discount, then the total cost must be less than the cost associated with a batch of 200 items, so:

$$1,080 + 10,100p < 100,800$$
$$10,100p < 99,720$$
$$p < \quad 9,873$$

So it would be reasonable to say that Tiger would order in batches of 1,000 if the price was £9.87 or less. This is a discount of $10 - 9.87/10 \times 100 = 1.3$ per cent.

13.5   Dealing with uncertainty

We have made a number of assumptions in dealing with this model, and two of these assumptions are highly unrealistic. One assumption made was there was no time interval between placing an order and receiving delivery of stock. We made this assumption to enable the firm to reorder when stock fell to zero and still avoid running out of stock. Now in practice there will be a time interval between placing an order and receiving a delivery, and we call this time interval the *lead time*. Let us suppose that the lead time for the delivery of headlamps is 3 days. Let us also suppose the Tiger Sports Cars work a 5 day week for 50 weeks in the year, i.e. a 250 working day year. We know that the annual demand for headlamps is 10,000, so the daily demand is for $10,000/250 = 40$ headlamps. Clearly, then, the firm will need $40 \times 3 = 120$ headlamps to tide it over while waiting for a delivery of stock. Previously, we advised the firm to place an order when stock fell to zero. In order to avoid a stockout, we must now advise that an order is placed when the level of stock falls to 120 headlamps.

A second rather unrealistic assumption was that demand was known and certain at 10,000 units per year ($= 40$ units per day). In practice, daily demand will vary randomly, and we will measure this degree of variation with the standard deviation. Suppose we assume that demand is normally distributed with an average of 40 units per day with a standard deviation of 5 units per day. Reordering when stock falls to 120 units will no longer guarantee that stockouts will be avoided because lead-time demand will now *average* 120 units during the lead time, and there will be a 50 per cent chance that *actual* lead time demand will exceed 120 units. So the firm will suffer a stockout on 50 per cent of occasions! What we need to ask is: what chance of a stockout would the firm be prepared to accept?

Suppose we know that the firm would be satisfied if a stockout was avoided on 99 per cent of occasions. Using our knowledge of normal distribution, we can represent the situation as shown in Fig. 13.5.

During the lead time, the demand for stock would only exceed X on 1 per cent of occasions. We can obtain the standard deviation for the lead-time demand that would form a sampling distribution of sample sums:

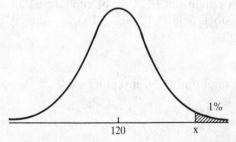

Figure 13.5

Standard deviation of lead-time demand $= \sqrt{5^2 + 5^2 + 5^2}$

$$= \sqrt{75}$$

$$= 8.66 \text{ units}$$

and the Z score would be

$$Z = \frac{X - 120}{8.66}$$

The normal distribution tables tell us that the appropriate value of the Z score in this case is 2.33, so

$$\frac{X - 120}{8.66} = 2.33$$

and $X = 120 + 2.33 \times 8.66 = 141$.

So, to ensure that stock would be sufficient on 99 per cent of occasions we would advise Tiger to reorder when stock level fell to 141 headlamps. This reorder level comprises 120 headlamps to meet average demand in the lead time and a *buffer stock* of 21 headlamps to meet random fluctuations in demand.

Finally, let us determine the effect of uncertain demand on inventory cost. We know that the firm will order batches of 200 headlamps, so on average it will place 10,000/200 = 50 orders per year. As each order will cost £8,

average annual ordering costs = 50 × £8 = £400.

On average, stock will be 21 units when a batch of 200 arrives, so on average

minimum stock = 21 units
maximum stock = 221 units

Average inventory will be (21 + 221)/2 = 121 units. As the annual holding cost is £4 per unit, average annual holding cost = £4 × 121 = £484. So the total annual inventory cost will be £400 + £484 = £884. Earlier, we calculated that if demand was certain at 10,000 units per year, then annual inventory cost would be £800, so the existence of uncertainty and the desire to be 99 per cent sure of avoiding a stockout has caused inventory costs to rise by £84 per year – a rise of 10.5 per cent.

13.6 Production run models – the economic batch quantity

Quite often a firm will not purchase stock from an outside supplier, but instead will produce the item itself and place it in store. The triangle representing the inventory cycle will no longer be right angled. Look at Fig. 13.6. It is assumed that the product is produced at a steady rate of R units per time period, and that stock is depleted at a steady rate of D units per time period. For a period of t_1 stock is accumulating because the rate of production exceeds the rate of demand. After this period t_1 production ceases, though stock continues to be depleted until the end of the inventory cycle t when stock falls to zero and production starts again. Working an example of this type from first principles is extremely complicated, and you would be well advised to learn the following formula which gives the optimum batch size q. Suppose R is the rate of production per time period, D is the demand per time period, both at a steady rate, cs is the cost of preparing the machinery for a production run, (called setup costs), and c_1 is the cost of holding a unit of inventory per time period, then:

$$q = \sqrt{\frac{2 \times R \times cs \times D}{c_1 \times (R - D)}}$$

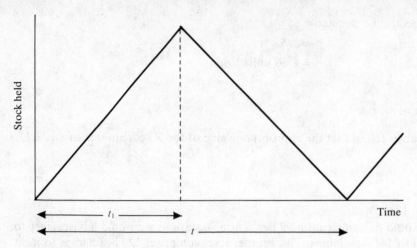

Figure 13.6

This quantity q is often called the economic batch quantity as opposed to the economic order quantity discussed earlier.

If the demand per time period is D, and the batch size produced is q, then the number of production runs made per time period is D/q and the setup cost per time period is csD/q. With the previous model, the batch size that minimised inventory costs equated ordering costs with holding costs; with this model the batch size that minimises inventory costs equates setup costs with holding costs. So the total inventory costs per time period is given by $2csD/q$.

Let us take an example. Suppose that a manufacturer can produce a certain product at a rate of 4,000 units per week. Demand for the product is 2,000 units per week and at a steady rate. It costs £50 to set up the machinery for a production run, and stock holding costs are 0.1p per unit per week. So here we have:

$$D = 2,000, \; R = 4,000, \; cs = £50 \text{ and } c_1 = £0.001$$

When a production run is started, the number of units produced to minimise inventory costs is:

$$\sqrt{\frac{2 \times R \times cs \times D}{c_1 \times (R - D)}}$$

$$= \sqrt{\frac{2 \times 4,000 \times 50 \times 2,000}{0.001 \times (4,000 - 2,000)}}$$

$$= 20,000$$

So the production run will stop after 20,000 units are produced. As the demand is 2,000 per week, then the batch would last for $20,000/2,000 = 10$ weeks, after which the next batch would be started. In other words, production would be on a ten-week cycle. The setup costs per time period are $cs \times D/q = 50 \times 2,000/20,000 = £5$ per week, and so the total inventory costs are £5 × 2 = £10 per week.

Review questions

13.1 A firm requires ten components per week, each component costing £15. It costs £1.20 to place an order and delivery is immediate. Warehousing costs are 75p per unit per annum and the cost of capital is 15 per cent per annum. Assuming a 50-week working year:

    (a) Find an expression for total annual inventory costs.
    (b) Find the batch size ordered to minimise inventory costs.
    (c) Find the length of the inventory cycle.

13.2 A firm requires 4,000 components per year, each component costing 40p. It costs £1 to place an order and warehouse costs are 10p per component per year. Cost of capital is 25 per cent.

    (a) Find the batch size that minimises inventory costs and the length of the inventory cycle.
    (b) Assuming the firm operates a five-day week for a 50-week year, and assuming that there is a two-day lag between orders and deliveries, how would this affect your answer to (a)?

13.3 A certain item costs £20 per unit and a firm requires 1,350 units per year. Ordering costs are £4 and warehouse costs 60p per unit per year. Cost of capital is 12 per cent per year. Advise the firm as to its optimal inventories policy.

13.4 For the firm in question 13.1, should it accept a 5 per cent discount for orders of 30 items?

13.5 For the firm in question 13.2, suppose it is offered a 1 per cent discount for orders of 500, and a 5 per cent discount for orders of 2,000; which, if either, offer should the firm accept?

13.6 For the firm in question 13.2, what is the minimum discount that must be offered to induce the firm to order batches of 1,000 items?

13.7 For the firm in question 13.2, suppose it is required to avoid a stockout with 99 per cent certainty. Find

    (a) the appropriate reorder level and
    (b) the effect that this has on total inventory cost
given that the standard deviation of daily demand was 5 components.

For the following three questions, find the run size to minimise total inventory costs, the length of the inventory cycle and the total inventory costs per time period.

13.8 Demand = 1,000 per week, production rate = 5,000 per week, setup costs = £100 and holding cost = 1p per unit per week.

13.9 Demand = 5,000 per month, production rate = 20,000 per month, setup costs £150 and holding cost = 30p per unit per month.

13.10 Demand = 1,500 per week, production rate = 4,500 per week, setup costs = £10 and holding costs = 20 per unit per week

14 Queuing and simulation

14.1 Introduction

Queues occur when a certain service facility is unable to keep up with the current level of demand. We are all familiar with this sort of situation: at certain times of day, for example, the checkouts at a supermarket are unable to cope with the rate of customer arrivals, and so customers have to wait for service. When we are going on our holidays we frequently meet situations when the demand for airspace exceeds the current availability, and we have long annoying waits at airport terminals. To us as consumers, queues are an irritating inconvenience, but to a firm the existence of queues can actually mean loss of business. Take the case of the supermarket, for example: if queues are allowed to become too long then they will spill over into the main body of the supermarket. Customers will be unable to obtain the goods they want from the fixtures and sales will be lost. Worse than this, if queues become too long then customers will take their business elsewhere! To take another example, if machines are 'queuing' for service because the service engineer is unable to cope with the demands made on his time, them a loss of productive output is bound to occur.

In this chapter, we will start by examining mathematically the theory of queues. However, at this level of study we are only able to examine very simple queuing situations if we stick to a rigorous approach. To examine more complicated queuing situations we must turn to a more *ad hoc* approach called *simulation*. As we shall see, this is a highly versatile technique applicable to a wide variety of situations.

14.2 Types of queuing situations

Perhaps the simplest way to describe a queuing situation is like this: individuals arrive at a service point, join the end of a queue (should one exist), receive service and then exit from the system. Diagrammatically, the situation would be represented like this:

ARRIVE → JOIN QUEUE → RECEIVE SERVICE → EXIT

The arrivals at an automatic car wash would be a typical example of such a situation. More complex situations can occur. There may be a single queue but multiple service points, and the head of the queue would go to any service point that happens to be vacant. An example of this is the queue discipline imposed in modern large post offices, and the situation can be represented diagrammatically like this:

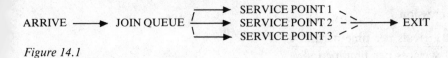

Figure 14.1

A third example is a multiple-queue–multiple-service point situation typically imposed in a supermarket. Diagrammatically, the situation would be represented like this:

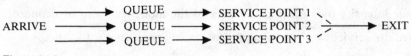

Figure 14.2

The first diagram describes the situation known as a *single channel* queue, whereas the second two diagrams describe what is known as a *multi-channel* situation. When examining queues mathematically, we shall confine our attention to the single channel situation.

So we see that one main classification of the queuing situation is by number of channels. A second classification is by queue discipline. By far the common situation you will find is the so-called first-come–first-served situation (that is, unless you happen to be waiting for a bus in Greece!). However, the situation can arise when an arrival into the system has a high priority, and will automatically go to the head of the queue. An example of this is the arrival of a case of cardiac arrest at the casualty department of a hospital.

14.3 A simple queuing model

We shall now examine a simple queuing situation often called the M/M/1 model. This model is dependent on certain assumptions.

(a) A single channel situation.
(b) Queue discipline is first-come–first-served. Each customer waits until served.
(c) The number of arrivals per time period can be described by a Poisson distribution.
(d) The number of services can be described by a negative exponential distribution.
(e) The service rate is greater than the arrival rate.

It is conventional to use certain symbols when examining queuing models. Within a given time period

λ is the average number arrivals into the system and
μ is the average number of services completed.

If the average number of arrivals into the system is λ, then it follows that the time interval between arrivals is $1/\lambda$. (If you don't understand this, substitute numbers for λ. If, say, eight arrivals occur in 1 hour, then the average time interval between arrivals is $\frac{1}{8}$ hour or 7.5 minutes.) Likewise, if μ is the average number of services per time period, then the average time taken to complete each service is $1/\mu$.

Suppose we wanted to determine the proportion of time that the system was utilised. This could be deduced from the expression

$$\frac{\text{Average number of arrivals per time period } \lambda}{\text{Average number of services per time period } \mu}$$

So if, for example, we can expect six arrivals per hour into a facility capable of performing eight services per hour, then we would expect the facility to be busy on $\frac{6}{8} = 0.75$, or 75 per cent of occasions. This fraction λ/μ is called the *traffic intensity*, and as it gives the proportion of time that the facility is busy, it must also give the probability of having to queue. Summarising, we have

1. Traffic intensity = probability of having to queue = λ/μ

Having determined the traffic intensity then other parameters of the model can also be deduced.

2. Probability of immediate service = $1 -$ (probability of having to queue) = $1 - \lambda/\mu$

What is not obvious, but can be proved mathematically is that

3. Average number of people in the system = $\lambda/(\mu - \lambda)$

By this is meant the number in the queue or being served.

4. Average time in the system = $1/(\lambda - \mu)$

and finally

5. Average time spent queuing = $\lambda/\mu(\mu - \lambda)$

Example

The spares department of a large motor vehicle repair shop has found it can deal with, on average, 80 requests for spare parts per hour. The arrival rate of mechanics needing spares averages 48 per hour. Find:

(a) the proportion of time that the spares department is busy,

(b) the probability that a mechanic receives immediate service,

(c) the average number of mechanics in the system,

(d) the average time spent in the system,

(e) the average time spent queuing.

In this case we have $\lambda = 48$, $\mu = 80$. Applying the above formulae:

(a) proportion spares department is busy = $\lambda/\mu = \frac{48}{80} = 0.6$,

(b) probability that mechanic receives immediate service = $1 - 0.6 = 0.4$,

(c) average number of mechanics in system = $48/(80 - 48) = 1.5$,

(d) average time spent in system = $1/(80 - 48) = \frac{1}{32}$ hour.

It has been estimated that every minute spent waiting for service costs the firm 10p in time lost by mechanics. It would be possible to employ a second person in the spares department to fetch the parts, while the first man does the paperwork. This would enable the spares department to deal with 120 mechanics per hour. The labour cost of employing the second man is estimated at £3 per hour. Would it be worth it employing a second man?

On average, 48 mechanics arrive in any one hour, spending 1/32 hours in the system. Hence, the total time spent waiting in any one hour $= 48 \times \frac{1}{32}$ hr $= \frac{3}{2}$ h $= 90$ min. The total cost of the waiting time will be $90 \times 0.1 = £9$. If the extra man is employed, then

average time spent in system $= 1/(120 - 48) = \frac{1}{72}$ h

The total waiting time in any hour would be $48 \times \frac{1}{72} \times 60 = 40$ min, and the total cost of waiting time is $40 \times 0.1 = £4$. Thus, the saving made by employing an extra man (£5 per hour) exceeds the hourly rate he would receive, and so the extra man should be employed.

14.4 Simulation

Have you noticed how someone latches on to a particular word or phrase, and then it is picked up by the mass media and hammered at us? War 'escalates', people 'commute' to work, law enforcement officers avoid 'no go' areas. We no longer have spares, we have back-up systems; no longer spoil the countryside, we pollute the environment. It would be easy to write a long list of such words and phrases. Nowadays, we are all familiar with the word 'simulation' – how did its popularity arise?

During the Second World War many raw materials were in short supply or not obtainable, and were produced artificially. Such goods were stamped 'imitation', and this proved to be a marketing disaster! – 'imitation' is a highly emotive word. It means made in the likeness of, but it did not mean this to potential purchasers! It meant a rather inferior substitute, to be avoided at all costs. If people were to buy such goods then the 'imitation' label would have to be dropped. Thus, 'imitation' silk became 'synthetic' silk, and sales were much healthier. Although the word synthetic was an improvement, it still did not produce the required product image. Now we have 'simulated' fur and 'simulated' suede. Why did the word 'simulated' succeed when 'imitation' and 'synthetic' failed? Probably because 'simulated' is more pleasant sounding, and almost certainly because fewer people knew what the word meant!

The exploration of space has certainly extended the popularity of the word simulation. To most people, it means a cartoon of an orbital docking or the re-entry of the capsule into the Earth's atmosphere. To NASA, simulation means a vital part of mission planning and mission control. At the Space Centre in Houston there is a mock-up of the space vehicles in which the astronauts familiarise themselves with layout and control systems. Clearly, it is better to learn in the simulator than to learn in the vehicle. The simulators were not merely used for astronaut training – if something goes wrong during the mission the failure can be reproduced on the simulators and experiments performed to rectify the fault. Again, it is better to experiment on the ground than in the vehicle.

The motives of simulation are clearly indicated by NASA. A problem is solved by simulation if it is too costly, too dangerous or just impractical to solve directly.

14.5 Monte Carlo methods

For the rest of this chapter, we will consider problems that are difficult to solve directly because they contain random elements. However, we will assume that the random elements can be quantified in an empirical probability distribution. We will examine how to apply simulation to queuing problems later, but initially we will take the case of a wholesale fishmonger selling boxes of fish. Daily demand will vary, but we will assume that the wholesaler has kept careful records of daily demand – records that can be summarised in an empirical probability distribution. Suppose he faces the following distribution of demand.

Number of boxes demanded per day	Probability
8	$\frac{1}{4}$
9	$\frac{1}{2}$
10	$\frac{1}{4}$

The wholesaler, for some reason, wishes to predict a typical chain of demand over the next 12 days. How can he do this?

The essence of simulation techniques is to substitute for the actual probability distribution some other distribution with the same probabilities. We can then use this second distribution to simulate the actual situation. Suppose we spin two coins: the probability distribution below would result:

Result	Probability
Head Head	$\frac{1}{4}$
Head Tail or Tail Head	$\frac{1}{2}$
Tail Tail	$\frac{1}{4}$

Now as this second distribution matches perfectly the distribution of demand, we can simulate a chain of daily demand by spinning coins. We could let two heads simulate a demand of 8 boxes, two tails simulate a demand of 10 boxes, and a head and a tail simulate a demand of 9 boxes. A prediction of demand for the next 12 days might look something like this:

Day	Result of spinning two coins	Simulated demand
1	H H	8
2	H H	8
3	H H	8
4	T T	10
5	H T	9
6	T H	9
7	H T	9
8	T T	10
9	H H	8
10	T T	10
11	H H	8
12	T H	9

You should realise that it is necessary to resort to simulation to predict a sequence of demands, because probability theory is a statement about long run frequency. We can use probability theory to state that, in the long run, we will experience a daily demand for 8 boxes on 25 per cent of occasions – we cannot use it to predict demand tomorrow, or on any particular day. Incidentally, you should notice that, using the probability distribution we would predict demand to be for 8 boxes on 25 per cent of occasions, whereas in the simulated sequence we experienced a demand for 8 boxes on $5/12 = 41$ per cent of occasions. This difference has arisen because we have taken a small sample of just 12 days. The longer the period over which we run the simulation, the closer would the simulated frequencies match the actual frequencies.

Clearly, it will not always be possible to spin coins and use the results as a basis for simulation. Suppose the pattern of demand facing the fishmonger was as follows:

Number of boxes demanded	Probability (%)
0	2
1	17
2	28
3	28
4	15
5	10
	100

Notice that in this case we have used percentage frequency as a measure of probability – it is often convenient to do this. We can use random numbers to obtain a frequency distribution that can form the basis of a simulation. Originally, random numbers were generated by a device rather like a roulette wheel (which explains why simulation is sometimes called Monte Carlo methods). Nowadays, random numbers are generated by computers. If you ask a computer to generate a string of random numbers for you, then you should not be able to detect any pattern in the string (after all, 'random' means 'without pattern'). Moreover, we can ask a computer to generate random numbers within certain defined limits (for example we would ask for random numbers in the range 1 to 10, or within the range 20 to 39.

Suppose we ask the computer to generate a random number in the range 00 to 99 inclusive. As there are 100 possibilities for the number generated, then the probability that the computer generates any particular number in this range is 1 per cent. The probability that the number is in the range 00–09 would be 10 per cent (note that 00–09 inclusive is ten numbers not nine). Likewise, the probability that the number generated is within the range 10–39 is 30 per cent. By grouping the random numbers in an appropriate fashion, we can obtain a distribution of random numbers that matches the distribution of demand. We can then use this distribution of random numbers as the basis of our simulation. The probability that the computer generates a random number in the range 00–01 is 2 per cent – the same as the probability that demand is zero. So if the computer generates a random number within the range 00–01, we will let this simulate a zero demand. The probability that the computer generates a random number in the range 02–18 is 17 per cent – the same as the probability that demand is for one box. So if the computer generates a random number within the range 02–18, we will let this simulate a demand for one box. The probability that two boxes are demanded is 28 per cent, and we can use the next 28 random numbers to simulate this, i.e. if the number generated by the computer is in the range 19–46. The complete distribution would look like this:

Random numbers	Frequency (%)	Simulates daily demand of
00–01	2	0
02–18	17	1
19–46	28	2
47–74	28	3
75–89	15	4
90–99	10	5

Suppose the computer produces the following list of random numbers:

27   89   99   97   03   95   31   50   91   34

Then reading from the table above we see that this simulates the following demand sequence:

2   4   5   5   1   5   2   3   5   2

We shall now extend the example of the fishmonger to illustrate how simulation can be used. Suppose the supply of fish also varies randomly with the following probabilities

Number of boxes supplied	0	1	2	3	4	5
Probability (%)	3	19	24	29	20	5

A box of fish costs the fishmonger £12, and the fishmonger sells a box for £23. Any box unsold at the end of the day is sold for cat food at £15 per box. If the fishmonger cannot supply a customer, then there is a loss of goodwill estimated at £5 per box. We will simulate 10 days trading and estimate average daily profit.

We have already simulated 10 days demand, so the first thing we must do is to simulate 10 days supply. To do this, we need a distribution of random numbers corresponding to the supply frequencies:

Random number	Frequency (%)	Simulates daily supply of
01–02	3	0
03–21	19	1
22–45	24	2
46–74	29	3
75–94	20	4
95–99	5	5

If the computer produces the following list of random numbers:

72   94   52   78   12   21   25   20   34   93

Then this simulates the following level of supply:

3   4   3   4   1   1   2   1   2   4

Summarising, we have used simulation to make the following predictions:

Day	1	2	3	4	5	6	7	8	9	10
Boxes supplied	3	4	3	4	1	1	2	1	2	4
Boxes demanded	2	4	5	5	1	5	2	3	5	2

Having simulated demand and supply, we have the necessary information to estimate average daily profit. You should notice the following rules must apply when estimating the numbers of unsold boxes and the levels of unsatisfied demand.

(1) The number of boxes sold on any day is the number demanded or the number supplied, whichever is the smaller. After all, the fishmonger cannot sell more boxes than he has bought!

(2) If, on any one day, the number of boxes supplied exceeds the number demanded, then the difference represents the quantity unsold.

(3) If, on any particular day, the number of boxes demanded exceeds the number supplied, then the difference represents the level of unsatisfied demand.

Day	Boxes supplied	Boxes demanded	Boxes sold	Boxes unsold	Unsatisfied demand
1	3	2	2	1	0
2	4	4	4	0	0
3	3	5	3	0	2
4	4	5	4	0	1
5	1	1	1	0	0
6	1	5	1	0	4
7	2	2	2	0	0
8	1	3	1	0	2
9	2	5	2	0	3
10	4	2	2	2	0
	25		22	3	12

Using the totals, we can deduce that

Revenue from sales to customers	= 22 × £23 =	£506
Revenue from sales for catfood	= 3 × £15 =	£45
		£551
Cost of boxes bought	= 25 × £12 =	£300
Loss of goodwill	= 12 × £5 =	£60
Total cost		£360
Total profit earned		£191
Average daily profit = 191/10		£19.10

14.6 A queuing model

The final stage in the manufacture of a certain product is inspection and rectification of faults. Previously, this was done by the distributor, but the firm making the product has decided that in future it will perform this operation itself. The production is scheduled such that one product leaves the production line every 20 minutes, but the actual time varies randomly like this:

Minutes late	3	4%
	2	6%
	1	18%
On time		36%
Minutes early	1	21%
	2	9%
	3	6%

Attempting to match production times, the inspector is supplied with sufficient mechanical aids to enable him to complete an inspection and to rectify any faults in 20 minutes, on the average. Some products will be relatively free from defects and will pass through inspection in less than 20 minutes, while others will take longer. The following distribution of inspection times was obtained:

Time	Frequency
18	9%
19	26%
20	32%
21	23%
22	10%

The factory operates for 8 hours each day, so 23 inspections are possible. The management wishes to derive various measures of performance for the proposed inspection system. In particular management wishes to know whether the 24 scheduled inspections can be completed on time. The management might also wish to know for how long during any day the inspector might be idle (which we will call idle time), and for how long products had to wait for inspection (which we will call waiting time). First, let us find the distribution of random numbers with the same frequencies as production times.

Minutes late	3	4%	00–03
	2	6%	04–09
	1	18%	10–27
On time		36%	28–63
Minutes early	1	21%	64–84
	2	9%	85–93
	3	6%	94–99

The production times give the times when the products arrive at the inspection shop. It will be logical to call them 'arrival times' and mark them with a minus when late, a plus when early, and zero when on time. Let us suppose that a product is waiting for inspection at the begining of the day (it will be the last product produced on the previous day). This means we will require 23 arrival times. The following random numbers were obtained from a computer:

72	94	52	78	12	21	25	20	34	93	27	89	99	97	03
95	31	50	91	91	34	46	73.							

These numbers simulate the following arrival times

+1	+3	0	+1	−1	−1	−1	−1	0	+2	−1	+2	+3	+3
−3	+3	0	0	+2	+2	0	0	+1					

If we note that the first arrival time refers to the second item produced (the first product is waiting inspection), and arrivals are scheduled at 20 minute intervals, actual arrival times are:

0	19	37	60	79	101	121	141	161	180	198	221	238	257
277	303	317	340	360	378	398	420	440	459				

Now we require a sequence of inspection times. The distribution of random numbers with the same frequency as the inspection times is:

Time	Frequency	Random pairs
18	9%	00–08
19	26%	09–34
20	32%	35–66
21	23%	67–89
22	10%	90–99

Obtaining twenty-four random numbers will give inspection times:

27	49	14	34	05	99	17	69	11	84	20	65	05	23	31
21	79	72	67	65	53	75	45	69						

So the inspection times will be (in minutes):

19	20	19	19	18	22	19	21	19	21	19	20	18	19	19
19	21	21	21	20	20	21	20	21						

There are four rules for this simulated system.

(1) The starting time for any inspection will be determined by either the finishing time of the previous inspection or the arrival time of the product; whichever is the later.

(2) The finishing time for any inspection will be the starting time plus the inspection time.

(3) If the arrival time of any product exceeds the time the previous inspection was finished, then the difference represents the inspector's idle time.

(4) If the time at which the inspection starts exceeds the arrival time, then the difference represents the product's waiting time.

The simulation shows that on this particular day, the inspector was idle for 11 minutes, and the 24 inspections took 7 minutes longer than scheduled. The average waiting time for products was $468/24 = 2.83$ min. This is tolerable as a queue is not forming.

Product number	Arrival time	Start	Inspection time	End	Idle time	Waiting time
1	0	0	19	19	0	0
2	19	19	20	39	0	0
3	37	39	19	58	0	2
4	60	60	19	79	2	0
5	79	79	18	97	0	0
6	101	101	22	123	4	0
7	121	123	19	142	0	2
8	141	142	21	163	0	1
9	161	163	19	182	0	2
10	180	182	21	203	0	2
11	198	203	19	222	0	5
12	221	222	20	242	0	1
13	238	242	18	260	0	4
14	257	260	19	279	0	3
15	277	279	19	298	0	2
16	303	303	19	322	5	0
17	317	322	21	343	0	5
18	340	343	21	364	0	3
19	360	364	21	385	0	4
20	378	385	20	405	0	7
21	398	405	20	425	0	7
22	420	425	21	446	0	5
23	440	446	20	466	0	6
24	459	466	21	487	0	7
					11	68

14.7 An inventory model

Let us use simulation to solve an inventory problem which is essentially similar to the inventory model considered earlier. A firm uses 200 components per day at a steady rate. Stock is replaced by purchasing from a supplier. It costs £10 to place an order. Stock holding costs are 0.1p per day. You should satisfy yourself that daily inventory costs would be minimised by ordering batches of 2,000 components every 10 days.

When dealing with inventories earlier, we assumed that the lead time (i.e. the time interval between placing an order and receiving delivery) was constant. We shall now consider the case where the lead time can vary randomly. Suppose the lead time is never less than 2 days nor more than 4 days, and has the following distribution:

Lead time	Frequency	Lead-time demand
2 days	29%	400 components
3 days	48%	600 components
4 days	23%	800 components

To cater for demand in the lead time, the firm has the choice of selecting three reorder levels: 400, 600 or 800. If a reorder level of 800 components is chosen (i.e. an order is placed when stock falls to 800) then no stockouts will occur. However, stock holding costs will be higher than with reorder levels of 400 or 600. If reorder levels of 400 of 600 used, then stockouts are inevitable. We shall simulate reorder policy such that an order is placed when the stock level is 600.

Before considering the simulated system, it will be useful to consider some preliminaries. Let us call the stock at the beginning of the cycle the opening stock, and suppose it is 2,000 components. This means that with a reorder level of 600 components, an order will be placed at the end of the seventh day. Now the length of the cycle will depend on the lead time. If the lead time is 2 days, the cycle length would be 9 days. Lead times of 3 or 4 days would give cycle lengths of 10 and 11 days respectively. Thus an opening stock of 2,000 components could give three different cycle lengths. Must the opening stock be 2,000 components? Suppose the opening stock is 2,000 and the cycle length is 9 days. When the new order arrives there will still be 200 components in stock, so the opening stock for the next cycle will be 2,200 components. With an opening stock of 2,200 components, stock will be reordered at the end of the 8th day and the cycle length would be 10, 11 or 12 days depending on the lead time. Thus there are six possible different cycles: a 2,000 or a 2,200 component opening stock each with three different cycle lengths.

The stock holding cost will be different for each of the six cycles. Suppose we consider the cycle which has an opening stock of 2,000, and a cycle length of 11 days. The average stock held on each of the 11 days would be:

Day	Average stock	
1	1,900	
2	1,700	
3	1,500	
4	1,300	
5	1,100	
6	900 ←	average
7	700	
8	500 ←	reorder
9	300	
10	100	
11	0	

It can be easily seen that the average stock per day over the cycle is 900 components. The holding cost over the cycle would be

$$900 \times 11 \times 0.1 = £9.90$$

Stock holding costs for other cycles could be calculated in a similar fashion.

Opening stock	Cycle length	Cost of cycle
2,000	9 days	£9.90
2,000	10 days	£10.00
2,000	11 days	£9.90
2,200	10 days	£12.00
2,200	11 days	£12.10
2,200	12 days	£12.00

How could a stockout occur? This will happen if the lead-time demand exceeds the reorder level. With a reorder level of 600, the lead-time demand would have to be 800 for a stockout to occur, i.e. if the lead time is 4 days. Now suppose the lead time was 2 days. With a reorder level of 600, lead-time demand would be 400 and there would be 200 units of inventory held at the end of the cycle (or the cycle has a closing stock of 200 components).

We can now simulate a reorder level of 600 components. The rules for the simulated system would be as follows:

(1) If opening stock is 2,000, reorder at the end of the 7th day, and at the end of the 8th day if the opening stock is 2,200.

(2) Generate a random number. The lead time is determined as follows:

Random digits	Probability	Record lead time of
00–28	29%	2 days
29–76	48%	3 days
77–99	23%	4 days

(3) Add the lead time to the reorder day to obtain cycle length.

(4) Enter the cost of the cycle from the table obtained earlier.

(5) If lead time is 3 days, record a zero closing stock and an opening stock of 2,000 for the next cycle.

(6) If the lead time is 4 days record a stockout. Also record a zero closing stock and a 2,000 opening stock for the next cycle.

(7) If the lead time is 2 days, record a closing stock of 200 and an opening stock of 2,200 for the next cycle.

If the opening stock for the first cycle is assumed to be 2,000 components, the simulation could be like this:

Cycle no	Opening stock	Random number	Lead time	Cycle length	Holding cost	Closing stock	Stockout
1	2,000	43	3	10	10	0	NO
2	2,000	40	3	10	10	0	NO
3	2,000	45	3	10	10	0	NO
4	2,000	86	4	11	9.9	0	YES
5	2,000	98	4	11	9.9	0	YES
6	2,000	03	2	9	9.9	200	NO
7	2,200	92	4	12	12	0	YES
8	2,000	18	2	9	9.9	200	NO
9	2,200	27	2	10	12	200	NO
10	2,200	46	3	11	12.1	0	NO
11	2,000	57	3	10	10	0	NO
12	2,000	99	4	11	9.9	0	YES
13	2,000	16	2	9	9.9	200	NO
14	2,200	96	4	12	12	0	YES
15	2,000	58	3	10	10	0	NO
16	2,000	30	3	10	10	0	NO
17	2,000	33	3	10	10	0	NO
18	2,000	72	3	10	10	0	NO
19	2,000	85	4	11	9.9	0	YES
20	2,000	22	2	9	9.9	200	NO
21	2,200	84	4	12	12	0	YES
22	2,000	64	3	10	10	0	NO
23	2,000	38	3	10	10	0	NO
24	2,000	56	3	10	10	0	NO
25	2,000	90	4	11	9.9	0	YES
				258	259.2		

Now 25 orders have been placed at a cost of £10 each. The ordering costs per day are:

$$250/258 = £0.968 \text{ per day}$$

The daily holding costs are:

$$259.2/258 = £1.005$$

Thus the simulated process gives a daily inventory cost of £1.973, with eight stockouts. Now let us compare this result with an 800 component reorder level, when stockouts could not occur. You should satisfy yourself that the simulation would be governed by the following rules.

(1) The opening stock could be 2,000, 2,200 or 2,400 giving reorders at the end of the 6th, 7th and 8th day respectively.

(2) There are nine possible cycles, the cost of which are:

Opening stock	Cycle lengths	Cost
2,000	8	£9.6
2,000	9	£9.9
2,000	10	£10
2,200	9	£11.7
2,200	10	£12
2,200	11	£12.1
2,400	10	£14
2,400	11	£14.3
2,400	12	£14.4

(3) A lead time of two days gives a closing stock of 400, and an opening stock of 2,400 for the next cycle. Lead times of 3 days and 4 days give closing stock of 200 and zero, and opening stocks for the next cycle of 2,200 and 2,000 respectively.

(4) The cycle lengths are obtained in the same way as the previous simulation.

Using the 800 component reorder level, an opening stock of 2,000 and the same lead times the simulation would look like this:

Cycle no	Opening stock	Lead time	Cycle length	Holding cost	Closing stock
1	2,000	3	9	9.9	200
2	2,200	3	10	12.0	200
3	2,200	3	10	12.0	200
4	2,200	4	11	12.1	0
5	2,000	4	10	10.0	0
6	2,000	2	8	9.6	400
7	2,400	4	12	14.4	0
8	2,000	2	8	9.6	400
9	2,400	2	10	14.0	400
10	2,400	3	11	14.3	200
11	2,200	3	10	12.0	200
12	2,200	4	11	12.1	0
13	2,000	2	8	9.6	400
14	2,400	4	12	14.4	0
15	2,000	3	9	9.9	200
16	2,200	3	10	12.0	200
17	2,200	3	10	12.0	200
18	2,200	3	10	12.0	200
19	2,200	4	11	12.1	0
20	2,000	2	8	9.6	400
21	2,400	4	12	14.4	0
22	2,000	3	9	9.9	200
23	2,200	3	10	12.0	200
24	2,200	3	10	12.0	200
25	2,200	4	11	12.1	0
			250	294.0	

Inventory costs are $250/250 + 294.0/250 = £2.176$ per day, an increase of £0.203 over the previous model. However, the previous model contained eight stockouts which must also be costed. Suppose that if a stockout occurs, the firm can send its own van to collect the 200 components required for that day. Collection will be on a cash-and-carry basis, and the firm will obtain a discount of £6. However, it costs £10 to send the van so the net cost of cash-and-carry would be £4. Hence, the stockout costs would be £4 plus the cost of holding one day's inventory, i.e. $£4 + 100 \times £0.001 = £4.10$. Thus the cost of the eight stockouts would be £32.80, which gives a daily stockout cost of $32.80/258 = £0.127$.

We can summarise the results like this:

Policy 1. Reorder level 800 components, inventory costs £2.176 daily.
Policy 2. Reorder level 600 components, inventory costs £2.100 daily.

On the basis of this simulation, policy 2 would be chosen.

Review questions

14.1 The service department of a charter airline is capable of repairing, on average, eight aircraft per week. The average rate of arrival of aircraft needing repair is five per week. Find

(a) the probability that an aircraft receives immediate attention;
(b) the average time an aircraft waits for repair;
(c) the average number of aircraft in the system.

14.2 Assume that for each day an aircraft is out of action the charter airline loses £2,000 in revenue. Extra staff could be employed to increase the service rate to 10 per week. What is the maximum the airline could afford to pay for the additional staff?

14.3 Look again at the fishmonger problem. Suppose that any boxes left over at the end of the day are put into cold storage for sale on the following day. Using the same data for supply and demand, find average daily profit and stock level at the end of the period.

14.4 This question refers to the production model analysed in the text. Suppose that instead of giving a distribution of deviations from scheduled production times, the problem had given actual production times. If the actual production times were:

Times (min)	17	18	19	20	21	22	23
Frequency	4	6	18	36	21	9	6

then very different results would have been obtained. Use the same random numbers as used in the text, and rework the simulation using actual production times.

14.5 Consider again the inventory model quoted in the text. Suppose policy 3 is to reorder when the stock falls to 400 components. Find the rules for the system. Using the same lead times as in the text, simulate 25 cycles and find the daily inventory costs. Compare the result with policy 2.

14.6 A National Health Service doctor decides to introduce an appointments system for daily consultations. A colleague supplies him with the following information as to patient punctuality:

Minutes early	3	6%
	2	29%
	1	41%
On time		12%
Minutes late	1	7%
	2	5%

The doctor times his consultations over a period, and derives the following frequency distribution:

12 minutes   10%
13 minutes   15%
14 minutes   28%
15 minutes   34%
16 minutes   13%

For convenience, he would like to issue appointments at 15 minute intervals. He wishes to have an idea of his idle time, the patients' waiting time, and whether he can complete his appointments on schedule. Simulate sixteen consultations and derive the required information.

Random numbers for arrivals
17   50   83   94   49   79   43   90   09   40   46   09   95   52   91   15
Random numbers for consultations
14   40   13   08   98   51   74   24   21   12   91   05   44   79   53   16

14.7   The distributions of arrivals and services at a supermarket checkout, per time period, are given below.

Arrivals	Frequency	Services	Frequency
8	8%	7	9%
9	22%	8	19%
10	38%	9	42%
11	26%	10	25%
12	6%	11	5%

Simulate 40 time periods and find the queue length.

Random numbers for arrivals
53   63   35   63   98   02   03   85   58   34
64   62   08   07   01   72   88   45   96   43
50   22   96   31   78   84   36   07   10   55
53   51   35   37   93   02   49   84   18   79
Random numbers for services
98   79   49   32   24   43   84   69   38   37
82   23   28   57   12   86   73   60   68   69
42   58   94   65   90   76   33   30   91   33
53   45   50   01   48   21   47   25   56   92

14.8   In the previous example, the queue will become very long. If the queue exceeds 10, it spills over into the selling area. Under such conditions, it is necessary to open extra checkouts. Find the appropriate time periods when it is necessary to do this.

14.9   A manufacturer uses a component at a rate of 50 per day. It costs £6 to place an order, and £0.015 to hold a unit of inventory for a day. Find the batch size q that would minimise inventory costs, on the assumption that delivery is immediate. Suppose that on 52 per cent of occasions delivery is on the day following the placing of the order, and on 48 per cent of occasions delivery is within 2 days. Simulate, and cost, 25 cycles for 50 and 100 reorder levels on the assumption that it costs the firm 50p per unit for each unit of stock short per day

Random numbers for lead times
78   93   79   30   96   95   90   13   52   53   90   58   09   21   13
61   31   86   59   32   75   99   59   44   65

14.10   In any one day, the price of stock traded on a certain stock exchange can rise, fall or remain unchanged. It is assumed that the direction of the change today depends upon what happened to the stock yesterday. The following probabilities are available:

		Today		
		Up	Unchanged	Down
Yesterday	Up	0.7	0.2	0.1
	Unchanged	0.3	0.4	0.3
	Down	0.1	0.3	0.6.

Moreover, if the price changes, it changes by the following amounts:

Amount of change (p)	Probability
0.5	0.30
1.0	0.40
1.5	0.20
2.0	0.05
2.5	0.05

Using the random digits below, simulate the next 10 days trading on the assumption that the stock closed yesterday unchanged at 100:

Random numbers for direction of change:
72   94   52   78   12   21   25   20   34   93
Random numbers for amount of change:
27   49   14   34   05   99   17   69   11   84

14.11 The distributions of daily supply and demand for a particular component was recorded as follows:

Daily demand	Frequency (%)	Daily supply	Frequency (%)
8	30	7	5
9	40	8	30
10	20	9	40
11	5	10	20
12	5	11	5

Simulate 10 days trading using the following random numbers:

For demand	21	52	68	33	31	99	91	72	56	90
For supply	94	95	66	78	91	08	52	26	47	01

Obtain a distribution of unsatisfied demand and find the stock level at the end of the period.

14.12 A car hire firm has four cars to rent. The daily distribution of demand for cars is as follows:

Number of cars demanded	Frequency (%)
0	10
1	20
2	20
3	25
4	15
5	10

The distribution of the time for which the car is demanded is:

Number of days demanded	Frequency (%)
1	15
2	25
3	30
4	20
5	10

Rental terms are £15 per day

Random numbers for number of cars demanded
38   53   29   32   91   52   48   89   70   24   06   31   48   09   06
60   24   42   08   08

Random numbers for number of days demanded
51   94   37   01   49   51   59   63   88   17   51   44   06   52   55
44   66   92   56   93   75   83

(a) Using the random numbers above, simulate the demand for vehicles over 20 days.
(b) Write a report advising the car hire firm on your conclusions from running this simulation.

15 Inequalities and linear programming

15.1 Introduction

In this chapter, we start by examining the meaning of an inequality in mathematical terms. You will be shown how inequalities are graphed, and what conclusions can be drawn from such graphs. Inequalities will then be related to the technique known as linear programming, and this technique will be used to solve product mix problems (i.e. which combination of products should be produced) and to solve blending problems.

15.2 Inequalities

If you come to think about it, one of the most important results of a system of numbers is the ability it gives us to range things in order of magnitude. We know that a man 185 cm tall is taller than a man 180 cm tall without ever having seen either of them. See how far you can get describing your girl friend (or boy friend) without the use of numbers and you will soon find that, however important the qualitative aspects, a really accurate description soon involves the use of numbers.

Moreover, the number system enables us to measure differences in magnitude. When we say that the man 185 cm tall is 5 cm taller than the man 180 cm tall, we are in fact making a number of mathematical statements such as:

180 cm + 5 cm = 185 cm	185 cm − 5 cm = 180 cm
180 cm is less than 185 cm	185 cm is greater than 180 cm

The difference in height is 5 cm; and so on.

Now for centuries mathematicians have concentrated on statements such as the first two above, which express the equality of two or more magnitudes. Perhaps this is because it is easier to visualise precisely things which are the same. But surely the important thing about measurement is the difference between one set of measurements and another. In the last 50 years or so a whole new branch of mathematics has been developed based on the fact that quantities are unequal, and it is this that we must first turn our attention to.

To do this you must first understand two basic signs:

(a) $>$, meaning greater than (e.g. $x > y$, x is greater than y) and its allied $\geqslant$, is greater than or equal to (e.g. $x \geqslant y$, x is greater than or equal to y).

(b) $<$, is less than (e.g. $y < x$, y is less than x). Can you see the meaning of $y \leqslant x$?

Suppose we say that $y < 2 < x$. We interpret this as y is less than 2, which is less than x, and can illustrate it in this way:

range of y values   range of x values

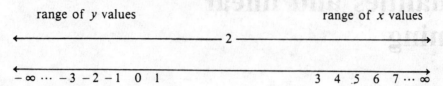

$-\infty \cdots -3 \; -2 \; -1 \quad 0 \quad 1 \qquad\qquad 3 \quad 4 \quad 5 \quad 6 \quad 7 \cdots \infty$

Now in this example neither x nor y is equal to 2, and so it should be apparent that y must always be less than x.

It is equally obvious to you that if $x > 2 > y$, then $x > y$. This is really saying the same thing in reverse but it expresses a very important feature of the 'greater than', 'less than' relationship: it is transitive.

15.2.1 Simple rules of inequalities

All other relationships are equally simple:

(a) Adding a number to both sides of an inequality does not change the sense of the inequality. Thus if $a > b$, $a + 2 > b + 2$; and if $a < b$, $a + 2 < b + 2$.

You will realise of course that this rule is also true if the number we add is a negative number. If $a > b$, $a + (-1) > b + (-1)$, i.e. $(a - 1) > (b - 1)$.

(b) Multiplying both sides of an inequality by the same positive number does not change the sense of the inequality. Thus if $c > d$, $2c > 2d$. More generally, if $c > d$ and $s > 0$, $cs > ds$.

(c) Multiplying both sides of an inequality by the same negative number reverses the sign of the inequality. Thus if $a > b$, $-2a < -2b$.

If you find this one confusing, try it with numbers rather than letters: $5 > 3$ but $-2 \times 5 < -2 \times 3$, i.e. $-10 < -6$.

This is all you need to know to go a long way in the mathematics of inequalities.

15.2.2 Simple linear inequalities or linear inequalities with a single variable

Example 1

Suppose we have an equation in the following form:

$2x + 4 > 3x + 2$

Subtracting $2x$ from both sides, we have:

$4 > 3x - 2x + 2$ or $4 > x + 2$

Subtracting 2 from both sides

$4 - 2 > x + 2 - 2$ or $2 > x$

More normally we would say $x < 2$.

Now there is no single value of x, but a range of numbers within which the value must fall; but nevertheless the inequality is solved and every value of x which is less than 2 satisfies it.

15.2.3 Graphing and inequality

Just as it is possible to draw the graph of an equation, so it is possible to draw the graph of an inequality.

15.3 Ordered pairs

Example 2

Consider the inequality $2x + y > 1$. It describes all combinations of x and y that satisfy the condition $2x + y > 1$.

We know already how to graph $2x + y = 1$. It will be the straight line joining the points $(\frac{1}{2}, 0)$ $(0, 1)$. Remember that every combination of values of x and y on this line will satisfy the equation $2x + y = 1$. See Fig. 15.1.

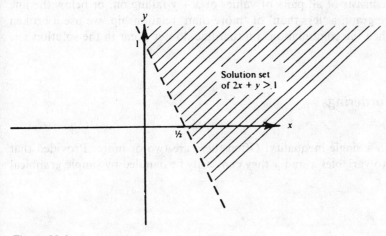

Solution set of $2x + y > 1$

Figure 15.1

But if the inequality $2x + y > 1$ is to be satisfied, either the value of x must be greater than the value indicated by the graph, or the value of y must. Thus any point lying above the line $2x + y = 1$ gives a pair of values of x and y which will satisfy $2x + y > 1$. We have many such pairs of values, and as a group we may call then the *solution set* of $2x + y > 1$.

Remember, however, that the solution set does not include any pair of points on the line $2x + y = 1$: only above it.

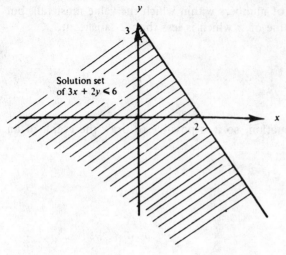

Figure 15.2

Example 3

Let us now depict geometrically the inequality $3x + 2y \leqslant 6$ (Fig. 15.2).

We proceed exactly as before, graphing the straight line $3x + 2y = 6$.

In this case, however, the value of $3x + 2y$ must be less than, or equal to 6 and this would be true if the value of x or y were to be less than, or equal to the value shown on the line. In other words, the solution set for $3x + 2y < 6$ consists of all pairs of values of $x + y$ falling on, or below the line $3x + 2y = 6$. Notice that when we graph a 'less than' or 'more than' relationship, we use a broken line. If a broken line is drawn, then the combinations on that line do not occur in the solution set.

15.4 Simultaneous linear orderings

It is seldom we have to deal with a single inequality. Often there are two or more. Provided that they do not involve more than two variables x and y they can easily be handled by simple graphical methods.

Example 4

Suppose we have a pair of linear inequalities:

$$x + y > 1$$
$$2x - y > 1$$

Figure 15.3 represents the two inequalities. If both are to be satisfied there must be an area where the solution sets coincide, i.e. pairs of values of x and y will satisfy both inequalities. The doubly-shaded area represents this coincidence of the solution sets.

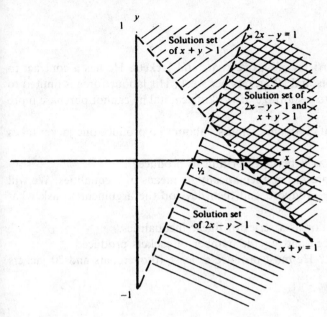

Figure 15.3

15.5 Linear expressions and linear programming

Those of you trained in traditional mathematics where the normal relationship is of the type $a = b$ or $2x + 4y + z = 26$ may be wondering why you have had to spend some time mastering expressions which are basically inequalities. Once you begin to try to apply mathematical techniques to problems of production, you will not wonder much longer. Most problems you will meet are of the type which involve some *constraints* as a result of having resources limited in quantity or with some limitation on quality attainable. The rest of this chapter is concerned with a simple application of the technique you have just learned to the type of problem the businessman is constantly facing. Possibly without realising it, you would have been studying a part of the technique known as 'linear programming'.

This expression is merely management jargon for the technique of reducing a practical problem to a series of linear expressions and then using those expressions to discover the optimal, or best, solution to achieve a given objective. That objective may be the maximisation of profit, the minimisation of cost, or possibly even the best proportions in which to blend Scotch whisky. Whatever the objective, the method tends to be the same.

Of course, not all industrial problems are amenable to these methods. There are times when it is quite impossible to express a given situation in linear form. For these, other and more complex methods of analysis must be used. But you would be surprised how many problems are capable of being solved by relatively simple techniques.

We will first concentrate on the simplest type of problem, those involving only two variables. The reason for this is that such problems are usually capable of simple graphical solution. You will readily appreciate that three variables involve three-dimensional graphs which few people really understand, and with more than three variables we enter the realm of graphs which rapidly become unmanageable.

Example 5

Suppose a small tailor is producing two articles only, overcoats and jackets. He has a contract to supply a department store with 10 overcoats and 20 jackets per week. His labour force is limited to five men working a 40-hour week and due to a shortage of working capital he cannot purchase more than 225 yards of cloth per week.

To produce one overcoat takes 5 yards of cloth and 5 hours' labour. To produce one jacket takes 3 yards of cloth and 2 hours' labour.

What combination of coats and jackets is it feasible for him to produce?

This is a typical problem that can be expressed mathematically by means of inequalities. We will first examine the *product-mix* that can be produced and then develop the argument to ask what product-mix should be produced.

First we state the problem facing the tailor as a series of linear inequalities.

Let x = the number of overcoats produced and y = the number of jackets produced.

How does the contract affect his output? He must produce at least 10 overcoats and 20 jackets to meet this contract. Hence:

$$x \geqslant 10 \text{ and } y \geqslant 20$$

(These are graphed in Fig. 15.4.)

There are, however, constraints imposed on the volume of production by the fact that he has available only 200 hours' labour and 225 yards of cloth. Let us consider labour first. If we produce x overcoats per week it will take $5x$ hours of labour. Similarly, y jackets will take $2y$ hours of labour.

But we have only 200 hours of labour available, so $5x + 2y$ cannot be greater than 200. Thus we

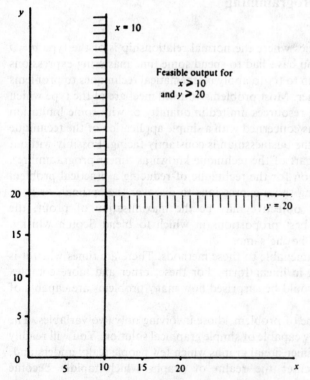

Figure 15.4   Contract constraints

194   **Inequalities and linear programming**

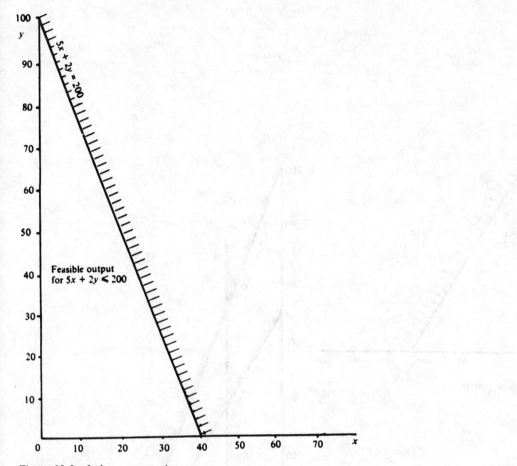

Figure 15.5   Labour constraint

may say:

$$5x + 2y \leqslant 200$$

(The labour constraint is drawn in Fig. 15.5.)

Considering now the cloth constraint, we can see that x overcoats take $5x$ yards of cloth, and y jackets take $3y$ yards of cloth. With only 225 yards available it is apparent that: $5x + 3y \leqslant 225$ (which is drawn in Fig. 15.6.)

We have now expressed mathematically the situation of the tailor. It is simply: if he produces x overcoats and y jackets:

$$x \geqslant 10 \qquad\qquad y \geqslant 20$$
$$5x + 2y \leqslant 200 \qquad 5x + 3y \leqslant 225$$

Before we can go further we must determine the combinations of x and y which are feasible, given the restrictions on labour and raw materials. By far the easiest way of doing this is to sketch the four inequalities, and by examining the solution sets determine what values of x and y satisfy all the inequalities – which we have done in Fig. 15.7.

In this case we have shaded the combinations of output which are not feasible. It is immediately apparent that there is a polygon ABCD which is unshaded. Within this area production is feasible

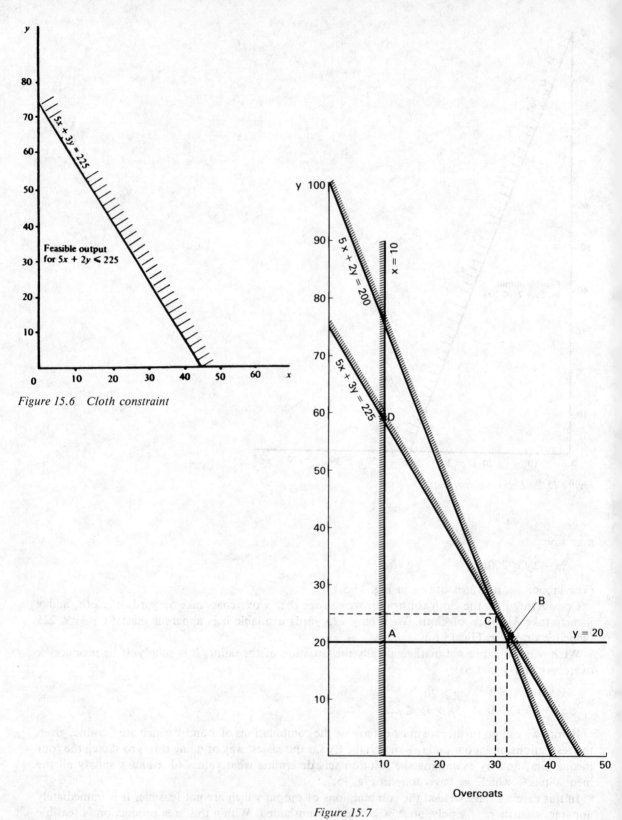

Figure 15.6   Cloth constraint

Figure 15.7

since neither the contract nor the limited labour force nor the restricted supply of cloth prevents it. This is why such an area is known as a feasibility polygon.

We have not yet of course solved any real problem for the tailor. All he has yet is a number of outputs of overcoats and jackets which he knows are possible. It could be 40 jackets and 20 overcoats, or 30 jackets and 25 overcoats. So does the feasibility polygon really help?

15.6 Profit maximisation – two-dimensional model

Think once again of the problem we have already partially examined. We have expressed the constraints on his output in the form of linear inequalities, viz.:

$$x \geqslant 10 \qquad\qquad y \geqslant 20$$
$$5x + 2y \leqslant 200 \qquad 5x + 3y \leqslant 225$$

You have learned how to graph and interpret expressions of this type and should understand how to obtain the feasibility polygon showing all possible combinations of jackets and overcoats the tailor can produce within the limits of the restrictions imposed on him by the availability of labour and working capital. (If you do not, turn back now before reading further.)

To use what we have obtained we must now consider the objective desired by the tailor. It is highly probable that an important objective will be to maximise profit and we will concentrate on this. The tailor knows that he makes a profit of £2 on each overcoat he sells and a profit of £1 on each jacket he sells. If, as before, he sells y jackets and x overcoats, we may express his total profit as profit = $y + 2x$ and the problem we are posing is what combination of jackets and overcoats will maximise this profit, and still satisfy the constraints. It is essentially the basic economic problem of how best to use limited resources which are capable of being used in more than one way.

We can state the problem this way (π is total profit):

Maximise
$$\pi = y + 2x$$
$$\text{Subject to } x \geqslant 10 \qquad 2y + 5x \leqslant 200$$
$$y \geqslant 20 \qquad 3y + 5x \leqslant 225$$

The expression 'maximise $\pi = y + 2x$' is called the objective function: it states in mathematical terms that the firm's objective is profit maximisation.

Referring again to Fig. 15.7, it should be obvious to you that combination B is preferable to combination A: both involve the production of 20 jackets, but as combination B produces more overcoats, then it must earn more profit. Likewise, combination D is preferred to combination A as it produces the same number of overcoats, but more jackets. In fact, the output that maximises profit must lie somewhere on the line D to C to B, and will be one of these combinations. Let us first examine combination B: reading from the graph, we can see that 20 jackets and 32 overcoats will be produced, and the profit earned from this combination will be £1 × 20 + £2 × 32 = £84. In a similar fashion, we can read off the other two combinations from the graph, and compute the appropriate level of profit:

Combination	Overcoats (x)	Jackets (y)	Profit
B	32	20	£84
C	30	25	£85
D	10	58	£78

So we can see from this table that in order to maximise weekly profit at £85, the tailor should produce 32 overcoats and 25 jackets.

Having solved the problem, two points are worthy of note here:

(a) The combination that maximises profit is on the intersection of the line $5x + 2y = 200$ (which is the labour constraint) and the line $5x + 3y = 225$ (which is the material constraint), so the combination that maximises the tailor's profit also maximises the use of inputs, as labour and materials are fully used.

(b) If you cannot read off the value from your graph accurately, then you should ascertain the appropriate equations at that point, and solve them simultaneously. For example, at combination D the lines $5x + 3y = 225$ and $x = 10$ intersect:

$$5x + 3y = 225 \tag{1}$$
$$x \qquad = 10 \tag{2}$$

Multiplying equation (2) by 5 and subtracting it from equation (1) gives:

$$\begin{array}{l} 5x + 3y = 225 \\ \underline{5x \qquad = 50} \\ 3y = 175 \\ y = 58.333 \end{array}$$

So the value we read off from the graph is wrong: the value of y is really 58.333 and the profit level at that output is:

$$10 \times £2 + 58.333 \times £1 = £78.333$$

Let this be a lesson to you! Once you have determined your optimum combination from the graph, check by solving the appropriate simultaneous equations.

15.7 Tips on drawing the diagrams

We will now look at a method that will quickly and efficiently solve two problems you may encounter when solving linear programming problems: how to scale the axes and how to draw the lines bounding the feasibility polygon. Always start off by constructing a table like this:

Constraint	Inequality	x	y
Labour	$5x + 2y \leqslant 200$	40	100
Material	$5x + 3y \leqslant 225$	45	75
Contract	$x \geqslant 10$	10	10
Contract	$y \geqslant 20$	20	20

The column headed x has been found by dividing the coefficient of x in the inequality into the constant (i.e. $200/5 = 40$, $225/5 = 45$, etc.). Now the maximum value in each column gives you the scaling for that axis. So the x axis would be scaled up to 45 and the y axis would be scaled up to 100. Moreover, the values in these columns can be used to draw the lines bounding the inequalities. So to draw the labour constraint, we would join 40 on the x axis with 100 on the y axis. You are strongly advised to use this technique.

15.8 Cost minimisation

One of the earliest applications of linear programming was the determination of the most economical use of raw materials to achieve a given objective.

Example 6

Let us suppose that a manufacturer of animal feeding stuffs is asked to supply 40 tons of feed per week to a large farmer. The farmer specifies that the food must contain a minimum of 25 per cent fat and 20 per cent protein, but that the total quantity may otherwise be made up by any bulk feed.

Two convenient raw materials are available, A and B. A contains 50 per cent fat and 25 per cent protein; B contains 25 per cent fat and 40 per cent protein.

The manufacturer has in stock 12 tons of A which he wishes to use, but further supplies are readily available. On the other hand, he is able to obtain only 20 tons of B per week. The price of both A and B is £25 per ton. In what proportions should he mix A and B in order to satisfy the farmer's requirements and minimise his cost of production?

The problem may be expressed mathematically as follows:

(a) $A \geqslant 12$ because of the need to use existing supplies.

(b) $B \leqslant 20$ because of the limitation on supplies. Also $B \geqslant 0$ as we cannot add a negative quantity.

(c) The desired fat content of the mixture is 10 tons (25 per cent of 40 tons) minimum, so 50 per cent of $A + 25$ per cent of $B \geqslant 25$ per cent of 40. That, is $2A + B \geqslant 40$.

(d) The desired protein content is 8 tons (20 per cent of 40 tons) minimum, so 25 per cent of $A + 40$ per cent of $B \geqslant 20$ per cent of 40. That is, $5A + 8B \geqslant 160$.

(e) The restriction on demand is 40 tons per week so $A + B \leqslant 40$.

(f) The cost of the mixture $C = 25A + 25B$.

Our full instructions read:
Minimise $C = 25A + 25B$ subject to

$$A \geqslant 12 \qquad 0 \leqslant B \leqslant 20 \qquad A + B \leqslant 40$$
$$2A + B \geqslant 40 \qquad 5A + 8B \geqslant 160$$

The only one of these constraints that might confuse you is the demand constant. Why do we say $A + B \leqslant 40$ rather than say $A + B = 40$? The point is that we may be able to meet the specifications of the feed by using quantities of A and B which total less than 40 tons and make good the difference by using a cheap edible bulk material rather than the more expensive raw materials.

Constructing our table:

Constraint	Inequality		A	B
Supply	A	$\geqslant 12$	12	
Supply	$B \leqslant 20$			20
Fat content	$2A + B \geqslant 40$		20	40
Protein content	$5A + 8B \geqslant 160$		32	20
Quantity	$A + B \leqslant 40$		40	40

So both axes will be scaled to 40 tons (Fig. 15.8).

Thus, the area bounded by PQRST shows the feasible region and, as we wish to minimise cost, it would seem reasonable to test the points nearest to the origin. We speculate that either P, Q or R is likely to be the cost-minimising combination. Now the only combination that we can feel confident in reading from the graph is combination P (32 tons of A and no tons of B), so we must use the simultaneous equation technique for the other two combinations.

Combination Q lies on the intersection of:

$$2A + B = 40 \tag{1}$$
$$5A + 8B = 160 \tag{2}$$

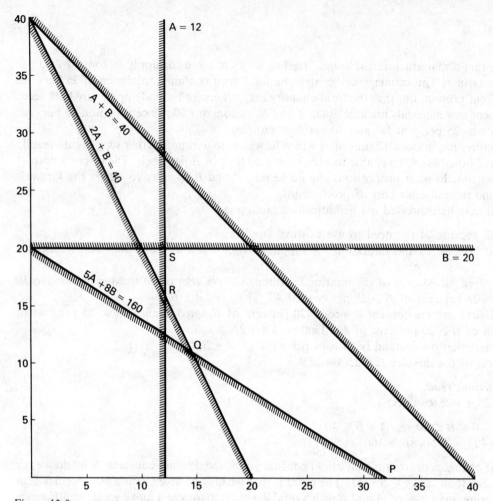

Figure 15.8

We can multiply equation (1) by 8 and subtract equation (2):

$$16A + 8B = 320$$
$$\underline{5A + 8B = 160}$$
$$11A \qquad = 160$$
$$A = 14\tfrac{6}{11}$$

and substituting this value in equation (1):

$$29\tfrac{1}{11} + B = 40$$
$$B = 10\tfrac{10}{11}$$

Combination R lies on the intersection of:

$$2A + B = 40 \tag{1}$$
$$A \quad = 12 \tag{2}$$

200   Inequalities and linear programming

Multiplying equation (2) by 2 and subtracting from equation (1) gives:

$$\begin{array}{rl} 2A + & B = 40 \\ 2A & = 24 \\ \hline & B = 16 \end{array}$$

We can now cost out these solutions:

Combination	Tons of A	Tons of B	Cost
P	32	0	£800
Q	$14\frac{6}{11}$	$10\frac{10}{11}$	£636.37
R	12	16	£700

So it is combination Q that would be chosen as this minimises cost. Because of the complexities of the numbers involved in this combination, it would seem feasible to suggest that 14.5 tons of A are mixed with 11 tons of B at a cost of £637.50. But if we do this, we must check that the fat constraint and the protein constraint are satisfied. For the fat constraint to be satisfied:

$$2A + B \geqslant 40$$
$$2A + B = 2 \times 14.5 + 11 = 40$$

so the fat constraint is satisfied. Turning now to the protein constraint:

$$5A + 8B \geqslant 160$$
$$5A + 8B = 5 \times 14.5 + 8 \times 11 = 160.5$$

and the protein constraint is satisfied.

Review questions

15.1   (a) If $x > 3$ and $y > 4$ find the greatest integer less than $3x + 4y$.
    (b) Depict geometrically the solution set of $-x + y > 1$. Would the solution set be different if $-x + y \geqslant 1$?

15.2   Sketch the solution set of the pair of inequalities:

$$x + y < 1$$
$$3x + 2y > 6$$

What would you conclude?

15.3   Sketch the solution set for the triple inequalities:

$$x + y \geqslant 2$$
$$x + 4y \leqslant 4$$
$$y > -1$$

15.4   Consider the non-linear inequality $x^2 - 6x - 16 < 0$. From this it follows that $(x + 2)(x - 8) < 0$. If $(x + 2) < 0$ what can you conclude about $(x - 8)$? Is your conclusion reasonable?
    If $(x + 2) > 0$ what can you conclude about $(x - 8)$? Is your conclusion reasonable?
    Deduce the solution to the inequality.

15.5 A firm is producing two brass ornaments, a standard and a de luxe model. The manufacturing process consists of two machines only:

	Machine 1	Machine 2
Standard model	1 hour	1 hour
Deluxe model	2 hours	5 hours

There are 20 hours available on machine 1 and 35 hours on machine 2. The firm makes a profit of 50p on the standard model and £1.50 on the de luxe model. Determine the output combination that would maximise profit.

15.6 A manufacturer is making two products, X and Y, with the following input requirements:

Product	Raw materials	Machine time	Labour
X	4 tons	2 hours	1 hour
Y	1 ton	1 hour	1 hour

Inputs are available in the following quantities per week:

90 tons of raw materials
50 hours of machine time
40 hours of labour

The profit on X is £4 per unit and the profit on Y is £3 per unit. What product mix maximises weekly profit?

15.7 An engineering firm specialises in converting standard trucks and vans into caravanettes. The conversion takes place in two workshops: the assembly shop where the structural alterations occur, and the finishing shop where the vehicles are fitted out. At present, the firm has a weekly capacity of 180 man-days in the assembly shop and 135 man-days in the finishing shop. Each truck requires 5 man-days' assembly, but each van only requires 2 man-days. Both vehicles require 3 man-days in the finishing shop. Each converted van earns £40 profit and each converted truck earns £200 profit. The firm has a contract to supply 10 converted trucks per week for export. Find the output combination that would maximise profit.

15.8 A farmer has two types of pig meal which he has to combine to produce a satisfactory diet for his pigs.
    A kilogram of Swillo contains 32 grams of nutrient A, 4 grams of nutrient B and 4 grams of C.
    A kilogram of Fillapig contains 16 grams of nutrient A, 8 grams of nutrient B and 20 grams of nutrient C.
    The pigs require at least 12 kilograms of pig meal each day, including a minimum of 256 grams of nutrient A, 56 grams of nutrient B and 80 grams of nutrient C. If a kilogram of Swillo costs £2 and a kilogram of Fillapig costs £3, find the least cost combination of pig meals.

16 The Simplex method

16.1 Introduction

In the previous chapter, we used graphical methods to solve product mix problems. However, graphical methods can only be used if we confine our attention to two products. In this chapter, we will examine a general method of solving linear programming problems called the *Simplex method* – a method that can be applied to any number of products. As a preliminary, we must first examine a method of solving simultaneous equations which is probably new to you. This method will then be applied to solving linear programming problems.

16.2 The detached coefficient method

If you think about it, there are two basic rules that apply to simultaneous equations.

Rule 1

If all the elements in an equation are multiplied or divided by a constant, then the equation will still hold.

Consider, for example, the equation

$$2x + 4y = 8.$$

We can multiply this equation through by (say) 3 to give

$$6x + 12y = 24$$

and this will not change the basis of the equation.

Rule 2

One can add (or subtract) any multiple of one equation from another without changing the basis of the equation.

These rules can often be applied to find the solution to simultaneous equations. For example, consider the equations

$$2x + 3y = 19 \qquad (1)$$
$$x + y = 8 \qquad (2)$$

Now subtracting twice equation (2) from equation (1) gives

$$0x + y = 3 \qquad (3)$$
$$x + y = 8 \qquad (4)$$

So from equation (3) we can conclude that $y = 3$. Now subtract equation (3) from equation (4)

$$0x + y = 3 \qquad (5)$$
$$x + 0y = 5 \qquad (6)$$

So from equation (6) we can conclude that $x = 5$. You should satisfy yourselves that $y = 3$ and $x = 5$ satisfies all the above equations.

One of the disadvantages of the traditional approach to solving simultaneous equations is that the variables x and y are repeated over and over again in the process of solution, yet it is the numbers that we are interested in. The approach we shall use here is to put the numbers into a *matrix or tableau*. Consider the simultaneous equations

$$3x + 2y + 4z = 19$$
$$2x + y + 2z = 10$$
$$2x + y + 4z = 16$$

In matrix form this would become

$$\begin{bmatrix} 3 & 2 & 4 & | & 19 \\ 2 & 1 & 2 & | & 10 \\ 2 & 1 & 4 & | & 16 \end{bmatrix}$$

The part of the matrix to the left of the vertical partition contains the coefficients of the variables, the first column containing the coefficients of x, the second the coefficients of y and the third the coefficients of z. To the right of the vertical partition is placed the quantities that each equation is equal to. In fact, you can regard the vertical partition as the matrix equivalent of the equals sign. The method we shall use to solve these equations is called the *detached coefficient method* and we shall now set out step-by-step instructions for using this method.

Step 1. Select any element in the coefficient matrix.

Step 2. Use Rule 1 to transform this element to one.

Step 3. Use Rule 2 to detach the remaining coefficients in that column, i.e. transform the remaining coefficients in the column to zero.

Step 4. Select another element but in a row or column not previously chosen. If all such elements equal zero then stop.

Step 5. Go to step 2.

We shall now apply these rules to the above equations.

$$\begin{bmatrix} 3 & 2 & 4 & | & 19 \\ 2 & 1 & 2 & | & 10 \\ 2 & ① & 4 & | & 16 \end{bmatrix}$$

The first step is to select any element in the coefficient matrix and transform it to 1. Well, the last two values in the second column are already one, so arbitrarily selecting the second of these (ringed) we must now transform the remaining elements in the second column to zero. We achieve this by

subtracting twice the third row from the top row and subtracting the third row from the second row

We now have

$$\begin{bmatrix} \text{\textcircled{-1}} & 0 & -4 & | & -13 \\ 0 & 0 & -2 & | & -6 \\ 2 & 1 & 4 & | & 16 \end{bmatrix}$$

We have transformed the second column, and the one is in the third row. We now choose another element not in this row or column and transform this to zero, and we have arbitrarily chosen the -1 in the first row (ringed). To change this -1 into 1 we divide the top row by -1. This gives

$$[1 \quad 0 \quad 4 \quad | \quad 13]$$

We now need to change the rest of the elements in the first column to zero. The second of these is already zero, so subtract twice the new first row from the third row to give

$$\begin{bmatrix} 1 & 0 & 4 & | & 13 \\ 0 & 0 & -2 & | & -6 \\ 0 & 1 & -4 & | & -10 \end{bmatrix}$$

We now must change the -2 in the second row into 1, as the second row and third column are the only ones not so far used. To do this we will divide the row by -2 to give

$$[0 \quad 0 \quad 1 \quad | \quad 3]$$

and if we subtract four times this row from the top row and add four times this row to the bottom row then the remaining elements in the third column will be zero.

$$\begin{bmatrix} 1 & 0 & 0 & | & 1 \\ 0 & 0 & 1 & | & 3 \\ 0 & 1 & 0 & | & 2 \end{bmatrix}$$

We have now solved the equations as we have

$1x + 0y + 0z = 1$ i.e. $x = 1$
$0x + 0y + 1z = 3$ i.e. $z = 3$
$0x + 1y + 0z = 2$ i.e. $y = 2$

Can you see the pattern that you must form to the left of the partition if you are to use this method to solve simultaneous equations? Each row and column must contain just one 1, the remaining elements being set to zero. So, if we have three equations, then there would be six possible patterns that could be formed in order to solve the equations:

```
1 0 0    1 0 0    0 1 0    0 0 1    0 1 0    0 0 1
0 1 0    0 0 1    1 0 0    1 0 0    0 0 1    0 1 0
0 0 1    0 1 0    0 0 1    0 1 0    1 0 0    1 0 0
```

16.3 Linear programming: Simplex method

Applying detached coefficients to linear programming is called the Simplex method. Let us start with a maximum profit, two product problem so that you can compare the Simplex method with the graphical approach. Suppose a manufacturer is making two products X and Y with the following input requirements:

Product	Raw materials	Machine time	Labour
X	4 tons	2 hours	1 hour
Y	1 ton	1 hour	1 hour

The inputs are available in the following weekly quantities:

90 tons of raw materials
50 hours of machine time
40 hours of labour

The profit is £4 on X and £3 on Y, and the objective is to maximise profit. Assuming the weekly

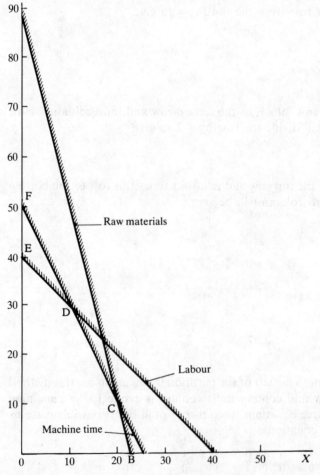

Figure 16.1

output is x units of X and y units of Y, the following model would describe the problem:

Maximise $\pi = 4x + 3y$

Subject to $\quad 4x + y \leqslant 90$
$$2x + y \leqslant 50$$
$$x + y \leqslant 40$$
$$x \geqslant 0$$
$$y \geqslant 0$$

Examining Fig. 16.1 we can deduce the following:

Combination	Quantity of X	Quantity of Y	Profit
B	22.5	0	£90
C	20	10	£110
D	10	30	£130
E	0	40	£120

So we would deduce that combination D is optimal, producing 10 units of X and 30 units of Y. Machine time and labour would be fully used, and there would be a surplus of raw materials of

$$90 - (4 \times 10 + 30) = 20 \text{ tons.}$$

The detached coefficient method cannot cope with inequalities, so we must transform the model into a system of equations. Consider the first inequality $4x + y \leqslant 90$, i.e. the raw material inequality. Suppose 10 units of each is produced, then the raw material requirement would be $4 \times 10 + 10 = 50$ tons, and there would be a surplus of 40 tons. What can we deduce about the size of the surplus? If we produce nothing, then the surplus is 90 tons, and if we use all of the raw material then the surplus would be zero. If the surplus is 'a' tons, then

$$0 \leqslant a \leqslant 90.$$

It follows from this that the amount used, plus the surplus, must equal 90 tons. Hence

$$4x + y + a = 90$$

The surplus 'a' is called a *slack variable*. If we call 'b' the surplus machine time and 'c' the surplus labour time, then we can rewrite the model as a system of equations

Maximise $\pi = 4x + 3y$

Subject to $\quad 4x + y + a = 90$
$$2x + y + b = 50$$
$$x + y + c = 40$$
$$x \geqslant 0, \quad y \geqslant 0$$
$$a \geqslant 0, \quad b \geqslant, \quad c \geqslant 0$$

Thus, we must satisfy the objective function subject to the constraint equations and subject to the additional constraint that all variables must be non-negative. We can represent the constraint equations in a partitioned matrix.

$$\begin{bmatrix} 4 & 1 & 1 & 0 & 0 & 90 \\ 2 & 1 & 0 & 1 & 0 & 50 \\ 1 & 1 & 0 & 0 & 1 & 40 \end{bmatrix}$$

$\uparrow$	$\uparrow$	$\uparrow$	$\uparrow$	$\uparrow$	$\uparrow$
x	y	a	b	c	Quantity

Stop for a moment and compare this problem with those considered in the last section. Previously, our problem had the same number of equations as they had variables. But with this problem, while we have 5 variables (x, y, a, b, c) we have only three equations. This difference is reflected in the coefficient matrix which is no longer square. We must conclude then, that this system of equations has no unique solution, but has many solutions. Which solution do we want? Clearly, the one that maximises profit.

Suppose for the moment we ignore the first two columns of this matrix, then we could, perhaps, treat it as a solution matrix. If it is legitimate to do this, then the solution would be

a = 90 (i.e. 90 tons of raw material are surplus)
b = 50 (i.e. 50 machine hours are surplus)
c = 40 (i.e. 40 hours of labour are surplus)

The solution would correspond to position A in Fig. 16.1. Now the implication of this solution is that no inputs are used — so if the solution is to be feasible it follows that nothing is produced and x and y must both equal zero. We are now in a position to formulate rules for interpreting such matrices:

(1) Identify the columns that form the solution matrix and read off the corresponding solution.
(2) Identify the remaining columns and set the corresponding variables at zero.

We have not yet entered the objective function into our matrix. To do this we use a *double partitioned matrix or Simplex tableau* and enter the objective function in the last row. As nothing is produced, it follows that profit is zero and so this is entered into the quantity column. Notice that the coefficients of x and y have been entered as negative quantitites. The reason for this will become apparent shortly, but for the time being take it to mean that as nothing is produced, the negative quantities in the last row measure profit forgone by not producing x and not producing y.

$$\begin{bmatrix} 4 & 1 & 1 & 0 & 0 & 90 \\ 2 & 1 & 0 & 1 & 0 & 50 \\ 1 & 1 & 0 & 0 & 1 & 40 \\ \hline -4 & -3 & 0 & 0 & 0 & 0 \end{bmatrix}$$
$$\begin{matrix} x & y & a & b & c & Q \\ & & \uparrow & \uparrow & \uparrow & \end{matrix}$$

Let us now deduce the effect of producing some units of one of the products. It would seem sensible to start with X, as it earns the greatest profit per unit. Suppose we decide to produce as much of X as possible — how many units can be produced? We can answer this problem by considering the x column and the Q column. The first row tells us it takes 4 tons of raw material to produce a unit of x, and 90 tons are available, i.e. there is sufficient raw material to produce $90 \div 4 = 22.5$ units. Reference to the second and third rows tells us there is sufficient machine time to produce $50 \div 2 = 25$ units, and sufficient labour time to produce $40 \div 1 = 40$ units. Raw material supplies restrict the output of x to a maximum of 22.5 units. But if we produce 22.5 units of X, we will use all the available raw material, and 'a' would be zero (remember, 'a' is the surplus raw material). Thus x is called the *entering variable*, (it enters the solution matrix) and 'a' the *departing variable* (it departs from the solution matrix). The non-zero variables (marked with an arrow in the matrix above) now become x, b and c.

We start as we did in the last section by dividing the top row by 4

$$[1 \quad \tfrac{1}{4} \quad \tfrac{1}{4} \quad 0 \quad 0 \mid 22\tfrac{1}{2}]$$

Now we must detach the remaining coefficients in the x column. The appropriate row transformations would be

Subtract twice the transformed top row from the second row
Subtract the transformed top row from the third row
Add four times the transformed top row to the bottom row:

After completing the transformations, the new matrix would look like this

$$
\begin{bmatrix}
1 & \frac{1}{4} & \frac{1}{4} & 0 & 0 & 22\frac{1}{2} \\
0 & \frac{1}{2} & -\frac{1}{2} & 1 & 0 & 5 \\
0 & \frac{3}{4} & -\frac{1}{4} & 0 & 1 & 17\frac{1}{2} \\
\hline
0 & -2 & 1 & 0 & 0 & 90
\end{bmatrix}
$$

$$
\begin{array}{cccccc}
x & y & a & b & c & Q \\
\uparrow & & & \uparrow & \uparrow &
\end{array}
$$

Again, we are interested only in the columns contained in the solution matrix (marked with an arrow). Reading-off this solution we have

$x = 22\frac{1}{2}$ (produce $22\frac{1}{2}$ units of X)
$b = 5$ (5 hours machine time surplus)
$c = 17\frac{1}{2}$ ($17\frac{1}{2}$ hours of labour surplus)
$a = 0$ (no surplus material)
$y = 0$ (Y is not produced)
$\pi = 90$ (Total profit is £90)

Let us again check the solution to ensure that it is feasible. As X earns £4 per unit profit, $22\frac{1}{2}$ units of X would earn £90. The raw material surplus is $90 - (4 \times 22\frac{1}{2}) = 0$, the machine time surplus is $50 - (2 \times 22\frac{1}{2}) = 5$ hours and the labour surplus is $40 - (1 \times 22\frac{1}{2}) = 17\frac{1}{2}$ hours. Once again, our solution is feasible and corresponds to point B on the graph.

Now let us produce some units of Y as well as X – how much can we produce? We must use the same method as before, i.e. divide the coefficients in the y column into the quantities in the Q column, and choose the smallest.

Why must we do this? If we do not choose the smallest value, then negative values will occur in the Q vector when we detach the remaining coefficients. You can verify this if you wish by multiplying the top row by 4 (or the third row by $\frac{4}{3}$) and detaching the remaining coefficients. We know that negative elements in the Q vector are not feasible. The second row gives the smallest total, so y is the entering variable, and b the departing variable. Multiplying the second row by 2 gives

$$[0 \quad 1 \quad -1 \quad 2 \quad 0 \mid 10]$$

To detach the remaining coefficients in the y column, we perform the following row transformations.

    Subtract one quarter of the transformed second row from the top row
    Subtract three quarters of the transformed second row from the third row
    Add twice the transformed second row to the bottom row.

$$
\begin{bmatrix}
1 & 0 & \frac{1}{2} & -\frac{1}{2} & 0 & 20 \\
0 & 1 & -1 & 2 & 0 & 10 \\
0 & 0 & \frac{1}{2} & -\frac{3}{2} & 1 & 10 \\
\hline
0 & 0 & -1 & 4 & 0 & 110
\end{bmatrix}
$$

$$
\begin{array}{cccccc}
x & y & a & b & c & Q \\
\uparrow & \uparrow & & \uparrow & &
\end{array}
$$

$x = 20$ (produce 20 units of x)
$y = 10$ (produce 10 units of y)
$c = 10$ (10 hours of labour surplus)
Profit is £110.
$a = 0$ (no surplus materials)
$b = 0$ (no surplus machine time)

This solution corresponds to position C on the graph.

Let us stop for a moment and consider exactly what we have been doing. We have selected an entering variable and performed row transformations to detach the remaining elements in the entering variable column. To detach the element in the objective function row, we have added some multiple of the entering variable row. As we have been adding, *this has caused profit to increase. This gives us a signal as to when we have reached maximum profit.*

If any column in the objective function is negative and we locate the entering variable in that column, then we must add to the objective row to detach the negative coefficient. Hence, profit must increase. If, on the other hand, the coefficient in the objective row and entering variable columns is positive, we will detach it by subtracting. This will cause profit to decrease. *Profit is at a maximum, then, when all the coefficients in the objective function cease to be negative.*

Now let us return to the matrix. If we look at the objective function row, we see that there is a negative coefficient in the '*a*' column, and if we locate the entering variable in this column we will increase profit. In this column, the second row gives the smallest total, but this cannot be chosen (why?). Instead, we take the third row, and '*c*' becomes the departing variable. Multiplying the third row by two gives

$$[0 \quad 0 \quad 1 \quad -3 \quad 2 \mid 20]$$

To detach the remaining coefficients in the '*a*' column we

Subtract half the transformed third row from the top row
Add the transformed row to the second and fourth row.

$$
\begin{bmatrix}
1 & 0 & 0 & 1 & -1 & 10 \\
0 & 1 & 0 & -1 & 2 & 30 \\
0 & 0 & 1 & -3 & 2 & 20 \\
\hline
0 & 0 & 0 & 1 & 2 & 130
\end{bmatrix}
$$

$$
\begin{array}{cccccc}
x & y & a & b & c & Q \\
\uparrow & \uparrow & \uparrow & & &
\end{array}
$$

$x = 10$ (produce 10 units of X)
$y = 30$ (produce 30 units of Y)
$a = 20$ (20 tons of raw material are surplus)
$b = 0$ (no surplus machine time)
$c = 0$ (no surplus labour)
The profit is £130.

As there are no negative coefficients in the objective row, this is an optimum solution: profit is at a maximum. You should notice that this corresponds with position D on the graph, and agrees with the graphical method.

Let us check that profit is at a maximum. If we made b the entering variable, can you see that x would have to be the departing variable? To detach the last element in the b column, we would have to subtract the top row from the bottom. Thus the bottom row would become

$$[-1 \quad 0 \quad 0 \quad 0 \quad 3 \mid 120]$$

Profit has declined by £10. In a similar fashion, we could show that if c was the incoming variable, then '*a*' would be the departing variable, and after detaching the coefficients in the c column profit would decline again by £10.

16.4 Lifting the constraints: scarcity values

So far we have been examining the problem facing the firm from the output viewpoint. We shall now concentrate on the problem from the input viewpoint, and in particular we will concentrate on the scarce inputs, labour and machine time. The reason for concentrating on the scarce inputs should be apparent to you – the firm does not use up all the raw material available to it. If more raw material was made available then this would simply add to the surplus, but if more labour or machine time was made available, then more could be produced and more profit earned.

Let us first examine the consequence of more labour being made available. In fact, we will assume that the firm can get hold of as much extra labour as it wants, so labour will no longer be a constraint. Look again at Fig. 16.1. If labour is no longer a constraint then we can imagine the diagram with the labour constraint removed. Combination F (which previously was not feasible) can now be produced. This combination would involve producing 50 units of Y and earning a profit of $50 \times 3 = £150$ – a rise of £20 over the previous optimal level. These 50 units of Y produced would require $50 \times 1 = 50$ hours of labour – a rise in labour utilisation of 10 hours over the previous optimal level. Thus an extra 10 hours of labour would add £20 to profit, an increase of £2 for each extra hour employed. This £2 per hour is called the *scarcity value* of labour. If the firm can increase its labour supply (perhaps by offering overtime work), then it should do so as long as its additional cost per hour does not exceed £2 – its contribution to profit. However, availability of machine time prevents more than 10 extra hours of labour being employed.

The same information that we have deduced from the graph could have been obtained directly from the solution matrix. Let us extract from the matrix column c (which refers to surplus labour) and the quantity column.

c	Q		
-1	10	←	production of x
2	30	←	production of y
2	20	←	surplus raw materials
2	130	←	profit

We have also identified the meaning of each value in the quantity column. Column c tells us what we should do if one extra hour of labour was made available to us. We should produce one less unit of X and two more units of Y. The raw material surplus would rise by 2 tons and profit would rise by £2. The reason that we produce less of X is to release machine time to enable more Y to be produced. If we wanted to compute the affect of an extra 10 hours of labour then we would multiply all the values in the c column by 10, i.e.

c	Q		
-10	10	←	production of x
20	30	←	production of y
20	20	←	surplus raw materials
20	130	←	profit

Comparing this new combination to the previous optimum, we can see that production of X would decline by 10 units (to zero), output of Y would rise by 20 units (to 50), the raw material surplus would increase by 20 tons (to 40 tons) and profit would rise by £20 (to £150). *If the supply of labour was increased by more than 10 hours no further change could take place as there is no longer an X output to eat into and release scarce machine time.* Finding the maximum by which a scarce resource can increase is called *sensitivity analysis*. To find this quantity involves locating any element in the column that is preceded by a minus sign. This element is divided into the corresponding quantity in the Q column and gives the maximum unit increase for the scarce resource. If there is more than one negative element, take the smaller result after division.

The other scarce resource in this example is machine time. This is given in column b.

b	Q		
1	10	←	production of x
-1	30	←	production of y
-3	20	←	surplus raw materials
1	130	←	profit

Interpreting this in a similar fashion, we would conclude that if an extra hour of machine time was made available, we should produce 1 more unit of X and one less unit of Y. The raw material surplus would decline by 3 tons and profit would rise by £1. To find the maximum increase to which this would apply we note that from the negative elements, $30/1 = 30$ and $20/3 = 6\frac{2}{3}$. Hence the maximum increase for machine time is $6\frac{2}{3}$ hours, which is imposed by the availability of raw materials.

The solution given in the quantity column is called the *prime solution*. It gives the combination of products that maximises the objective function (in this case profit). The solution obtained from the objective function row is called the *dual solution*. In this case, it gives the smallest valuation of the scarce resources that can account for all the profits of the output. In this case, we see that labour has a scarcity value of £2 per hour, and as 40 hours are available the total scarcity value of labour is $40 \times £2 = £80$. Likewise, the scarcity value of machine time is £1 per hour, and as 50 hours are available the total scarcity value of machine time is $50 \times £1 = £50$. Hence, the total value of the scarce resources is $£80 + £50 = £130$: the same as the total profit earned.

16.5   A mixed inequalities example

An engineering firm specialises in converting standard trucks and standard vans into mobile caravans. The conversion takes place in two workshops: the assembly shop where structural conversion takes place and the finishing shop where the vehicle is fitted out. At present, the firm has a weekly capacity of 180 man-days in the assembly shop and 135 man-days in the finishing shop. Each truck needs 5 man-days of assembly, but each van only requires 2 man-days. Both vehicles require 3 man-days of finishing. Each converted truck earns £200 profit and each converted van earns £40 profit. The firm has a contract to supply 10 converted vans each week for export. The objective of the firm is to maximise its weekly profit.

We can represent the problem mathematically like this:

Maximise $\pi = 200t + 40v$

Subject to
$$5t + 2v \leqslant 180$$
$$3t + 3v \leqslant 135$$
$$v \geqslant 10$$
$$t \geqslant 0$$

Figure 16.2 is a diagrammatic representation of the situation from which we can deduce:

Combination	Trucks	Vans	Profit (£)
B	0	10	400
C	32	10	6,800
D	30	15	6,600
E	0	45	1,800

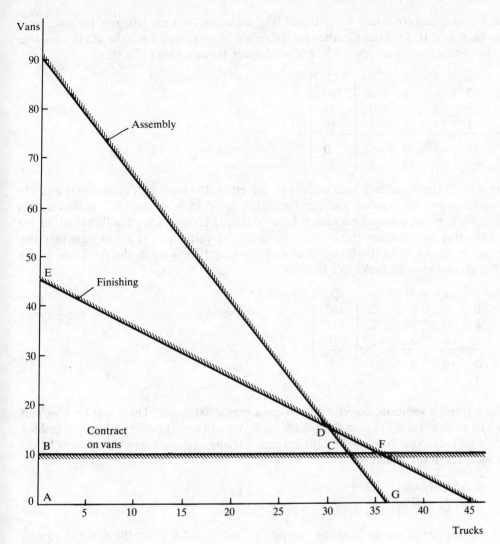

Figure 16.2

So we see that the firm would maximise its profit by converting 32 trucks and 10 vans. Assembly time would be fully used, but there would be a surplus of $135 - (32 \times 3 + 10 \times 3) = 9$ man-days finishing.

Let us now see how we would solve this problem using the Simplex method. We must again convert the inequalities to equations. Converting the first two constraints is no problem

$$5t + 2v + a = 180$$
$$3t + 3v + b = 135$$

where a and b are slack variables representing surplus assembly and finishing time. How can we deal with the contract constraint which is a 'greater than or equal to' ordering? Clearly, it would not make sense to add a slack variable, we must instead subtract a variable:

$$v - c = 10$$

It must be noted that the minus sign does not mean that c is negative. It means that c must be subtracted from v to satisfy the contract constraint: c is itself non-negative, and is called an *artificial variable.* What meaning can we attach to c? It will be the amount of vans left over for sale, once the contract has been met. If, for example, the firm converted 30 vans, then $c = 20$ would be available for sale after the contract has been met. The initial Simplex tableau looks like this:

$$
\begin{bmatrix}
\begin{array}{ccccc|c}
t & v & a & b & c & Q \\
5 & 2 & 1 & 0 & 0 & 180 \\
3 & 3 & 0 & 1 & 0 & 135 \\
0 & 1 & 0 & 0 & -1 & 10 \\
\hline
-200 & -40 & 0 & 0 & 0 & 0 \\
& \uparrow & \uparrow & & &
\end{array}
\end{bmatrix}
$$

This solution represents the situation where nothing is converted, 180 man-days of assembly and 135 man-days of finishing time are surplus, and corresponds to position A in Fig. 16.2. However, this solution is not feasible, as the contract on vans is being violated. To produce a feasible solution, we will make v the entering variable, but there will be no departing variable. The transformations are: subtract twice the third row from the first and second rows, add forty times the third row to the bottom row. The second tableau looks like this:

$$
\begin{bmatrix}
\begin{array}{ccccc|c}
t & v & a & b & c & Q \\
5 & 0 & 1 & 0 & 2 & 160 \\
3 & 0 & 0 & 1 & 3 & 115 \\
0 & 1 & 0 & 0 & -1 & 10 \\
\hline
-200 & 0 & 0 & 0 & -40 & 400 \\
& \uparrow & \uparrow & \uparrow & &
\end{array}
\end{bmatrix}
$$

We now have a feasible solution: convert 10 vans and earn £400 profit. There will be a surplus of 160 man-days of assembly and 115 man days of finishing. We can now proceed in the usual fashion and obtain an optimal solution. If we make t the entering variable, can you seen that a must be the departing variable? The transformations are:

divide the top row by 5 giving:

$$[1 \quad 0 \quad \tfrac{1}{5} \quad 0 \quad \tfrac{2}{5} \mid 32]$$

subtract three times the new top row from the second row and add 200 times the new top row to the bottom row:

$$
\begin{bmatrix}
\begin{array}{ccccc|c}
t & v & a & b & c & Q \\
1 & 0 & \tfrac{1}{5} & 0 & \tfrac{2}{5} & 32 \\
0 & 0 & -\tfrac{3}{5} & 1 & \tfrac{9}{5} & 9 \\
0 & 1 & 0 & 0 & -1 & 10 \\
\hline
0 & 0 & 40 & 0 & 40 & 6,800 \\
\uparrow & \uparrow & & \uparrow & &
\end{array}
\end{bmatrix}
$$

As all the coefficients in the last row are positive, this is the optimal solution. The firm should convert 32 trucks and 10 vans, earning a weekly profit of £6,800. This corresponds to position D on the diagram. Reading off the dual solution (from column a), we see that the scarcity value of assembly time is £40 per man-day. So, if the firm can contract-out some assembly time it should do so, providing the additional cost does not exceed £40 per man-day. Moreover, it should contract-out $9 \div \tfrac{3}{5} = 15$ man days. This would enable the firm to convert an extra $\tfrac{1}{5} \times 15 = 3$ trucks and increase profit by £40 × 15 = £600. All the finishing time would then be used. This situation corresponds to position F on the graph.

Shadow prices

When we considered scarcity values earlier, we intimated that the total value of scarce resources should equal the profit earned. Now in this example we have just one scarce resource – assembly time – which has a scarcity value of £40 per man-day. As 180 man-days are available, the potential profit from fully using this scarce resource is £40 × 180 = £7,200. But we see from the tableau that the firm earns only £6,800. How can we account for this £400 shortfall?

Look again at the optimal solution. The firm converts just 10 vans – just the right number to satisfy his export contract. In other words, the contract seems to be restricting his ability to increase his profit. Suppose the contract was removed, then a glance at Fig. 16.2 will show that combination G would become feasible. He could convert 36 trucks and no vans and earn a weekly profit of 36 × £200 = £7,200. This same information can be obtained from column c of the solution matrix.

c	Q		
$\frac{2}{5}$	32	←	converted trucks
$\frac{9}{5}$	9	←	surplus finishing
-1	10	←	converted vans
40	6,800	←	profit

Column c shows us what would happen if we could convert one van less. This would release sufficient scarce resources to convert $\frac{2}{5}$ths of a truck, the finishing time surplus would rise by 1.4 man-days and profit would rise by £40. Dividing the 1 in the c column into the 10 in the Q column gives us the limit (10) to this sensitivity analysis, i.e. if the contract was removed then we should decrease van conversions by 10 and in fact not convert any vans.

So, we would advise the firm to cancel its contract to convert vans as soon as possible. However, there is an alternative approach that we can take. The final figure in the c column tells us that every van that is converted reduces the profit potential by £40. We can regard this figure as a sort of *shadow price* – the extent to which vans are underpriced. If the price of converted vans (and hence the profit earned from them) was increased by this amount to £80 then the full profit potential could be realised. The profit then earned would be

$$32 \times 200 + 10 \times 80 = £7,200.$$

Finally, then, we would advise the firm that if it cannot negotiate an increase in price of £40 for converted vans, it should cancel its export contract as soon as possible.

A minimising example

The Simplex method is basically a maximising method, and to solve minimising problems some adjustments must be made. We gave a hint of how this can be done when considering scarcity values. Do you remember tht we found the *smallest* valuation of the scarce inputs that could account for all the profits of the outputs? It is a cardinal fact with the Simplex method, that if you maximise the primal problem then you will minimise the dual. So, if we can write the dual of a problem and maximise, then using the Simplex method we will minimise the primal problem. Let us consider an example.

An oil company puts additives into petrol to give improved performance and reduce engine wear. Ideally, each 10,000 litres should contain at least 40 mg of additive P, 14 mg of additive Q and 18 mg of additive R. The company can obtain stocks of ingredients which have the following weights of additive per litre:

Ingredient	mg of P	mg of Q	mg of R
X	4	2	3
Y	5	1	1

Both additives cost £1,000 per litre. The company wants to know the quantity of each ingredient that should be added to each 10,000 litres of petrol to satisfy the additive requirements at minimum cost. (We can assume that the quantity of each ingredient added will be so small that the total volume of petrol will be unaffected.) The primal problem is

$$\text{Minimise } 1{,}000x + 1{,}000y$$
$$\text{Subject to} \quad 4x + 5y \geqslant 40$$
$$2x + y \geqslant 14$$
$$3x + y \geqslant 18$$
$$x \geqslant 0$$
$$y \geqslant 0$$

The feasible region is shown in Fig. 16.3, and costing out each combination:

Combination	Litres of X	Litres of Y	Cost
A	0	18	£18,000
B	4	6	£10,000
C	5	4	£9,000
D	10	0	£10,000

So in order to minimise cost, the company should add 5 litres of ingredient X and 4 litres of Y to each 10,000 litres of petrol. The petrol would contain exactly the minimum requirements of additive P and Q, and the quantity of R would be

$$3 \times 5 + 4 - 18 = 1 \text{ mg}$$

in excess of the minimum requirement.

In order to solve this problem using the Simplex method, we will first assign a value of each mg of the additives. We will assign a value of £p to each mg of P, £q to each mg of Q and £r to each mg of R. So, the value of additives in each litre of ingredient X would be

$$4p + 2q + 3r$$

Now the company uses this ingredient solely because it contains additives. Hence, the company will certainly not wish the value of additives in a litre of ingredient X to exceed its cost, so

$$4p + 2q + 3r \leqslant 1{,}000$$

Likewise for ingredient Y

$$5p + q + r \leqslant 1{,}000$$

The total cost of ingredients in 10,000 litres of petrol would be

$$40p + 14q + 18r$$

Moreover, what we shall do is to find the largest valuation on additives that will account for the costs of the ingredients, i.e.

$$\text{Maximise } 40p + 14q + 18r$$
$$\text{Subject to } 4p + 2q + 3r \leqslant 1{,}000$$
$$5p + q + r \leqslant 1{,}000$$
$$p \geqslant 0, \qquad q \geqslant 0, \qquad r \geqslant 0$$

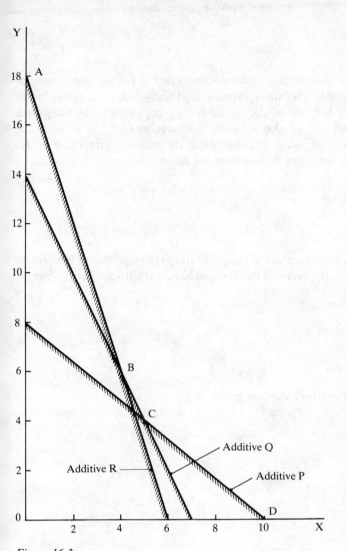

Figure 16.3

We have now deduced the (maximising) dual problem that is the mirror image of the (minimising) primal problem. Let us set out both problems side-by-side and compare them.

Primal Problem

Minimise $1,000x + 1,000y$

Subject to
$$4x + 5y \geqslant 40$$
$$2x + y \geqslant 14$$
$$3x + y \geqslant 18$$
$$x \leqslant 0, \quad y \geqslant 0$$

Dual Problem

Maximise $40p + 14q + 18r$

Subject to
$$4p + 2q + 3r \leqslant 1,000$$
$$5p + q + r < 1,000$$
$$p \geqslant 0, q \geqslant 0, r \geqslant 0$$

We see clearly that the minimising problem has been turned 'inside out'. What is a column in one becomes a row in the other, the inequality signs are reversed, the coefficients in quantity column of one problem become the coefficients in the objective function in the other. Only the non-negative conditions for the variables remain unchanged.

We will now prepare the dual problem for a Simplex solution. Adding a slack variable to the first constraint we have

$$4p + 2q + 3r + x = 1,000$$

We have chosen x as the slack variable to indicate that the equation refers to the value of a litre of ingredient X. What meaning can we attach to this slack variable x? Ideally, we want the value of additives in each ingredient to account for the cost of the ingredient. If they cannot, then the slack variable x will be a measure of the shortfall of the value of additives against the cost of the ingredient. It will be the opportunity cost of using this ingredient instead of using some other. Likewise, if we add a slack variable y to the second constraint we have:

Maximise $\quad 40p + 14q + 18r$
Subject to $\quad 4p + 2q + 3r + x = 1,000$
$\qquad\qquad\quad 5p + q + r + y = 1,000$
$\qquad p \geqslant 0, \qquad q \geqslant 0, \qquad r \geqslant 0$

Notice that for the first time we have *more than two unknown quantities* in the equations. We cannot graph this situation but, as we shall now see, the Simplex method can handle it without any problems. The initial Simplex tableau would be:

p	q	r	x	y	Q
4	2	3	1	0	1,000
5	1	1	0	1	1,000
-40	-14	-18	0	0	0

and performing row transformations in the usual way we have:

p	q	r	x	y	Q
0	$\frac{6}{5}$	$\frac{11}{5}$	1	$-\frac{4}{5}$	200
1	$\frac{1}{5}$	$\frac{1}{5}$	0	$\frac{1}{5}$	200
0	-6	-10	0	8	8,000

p	q	r	x	y	Q
0	1	$\frac{11}{6}$	$\frac{5}{6}$	$-\frac{2}{3}$	166.67
1	0	$-\frac{1}{6}$	$-\frac{1}{6}$	$\frac{1}{3}$	166.67
0	0	1	5	4	9,000.00

As all the coefficients on the objective function are positive, we have maximised the primal problem. However, we are basically interested in the minimising dual problem. Moreover, we saw earlier that to use the Simplex method we turned the problem inside-out. So, the quantities we want will now appear in the objective function row rather than in the quantity column. Reading off from this column we have $x = 5$ and $y = 4$ (i.e. add 5 litres of X and 4 litres of Y). Total cost = £9,000, and $r = 1$ showing that the mixture contains 1 mg of R in excess of the minimum requirements. This agrees precisely with the solution we obtained using graphical methods.

When we maximised, we performed a sensitivity analysis on the constraints by examining the columns of the solution matrix. As we have turned the problem inside-out, we will perform a sensitivity analysis on the rows of the solution matrix. The quantity in the first row clearly tells us something about Q. In fact the whole row tells us what we should do if the additive Q requirement was increased by 1 mg. Reading from this row we see that for each unit increase in the additive Q requirement we should add $\frac{5}{6}$ litre more of X and $\frac{2}{3}$ litre less of Y. This would meet the extra requirements at a *minimum cost increase* of £166.67 per unit increase of additive Q. The excess of additive R over the minimum requirements would rise by $\frac{11}{6}$ mg. It would be useful to perform a feasibility check to ensure that what is predicted by this matrix would in fact happen. We firstly reproduce the table giving the specification of each additive

Ingredient	mg of P	mg of Q	mg of R	Cost
X	4	2	3	£1,000
Y	5	1	1	£1,000

Let us first compute the effect of adding $\frac{5}{6}$ litre more of X:

mg of P	mg of Q	mg of R	Cost
$\frac{5}{6} \times 4$	$\frac{5}{6} \times 2$	$\frac{5}{6} \times 3$	$\frac{5}{6} \times £1,000$
$\frac{20}{6}$	$\frac{10}{6}$	$\frac{15}{6}$	5,000/6

Now we compute the cost of adding $\frac{2}{3}$ $(=\frac{4}{6})$ less of Y:

mg of P	mg of Q	mg of R	Cost
$-\frac{4}{6} \times 5$	$-\frac{4}{6} \times 1$	$-\frac{4}{6} \times 1$	$-\frac{4}{6} \times £1,000$
$-\frac{20}{6}$	$-\frac{4}{6}$	$-\frac{4}{6}$	$-4,000/6$

So the net affect of this change is:

mg of P	mg of Q	mg of R	Cost
0	1	$\frac{11}{6}$	166.67

Which agrees precisely with the predictions made in the first row. This analysis would hold for increases of up to $4 \div \frac{2}{3} = 6$ mg of Q.

In the same way, we can deduce that for each increase in additive P requirement, we should add $\frac{1}{3}$ litre more of Y and $\frac{1}{6}$ litre less of X. This would cause cost to rise by £166.67 and the surplus in additive R to fall by $\frac{1}{6}$ mg. This analysis would apply for increases in the additive of up to $1 \div \frac{1}{6} = 6$ mg of P, whereupon the additive R requirement would just be met. You should perform a feasibility check on this row to convince yourself that the predictions made by this row are true.

Review questions

Use the Simplex method in each case

16.1 A manufacturer is prepared to allocate up to 60 hours of labour, 14 machine hours, and 36 kg of raw materials per week for the production of two products, X and Y. Product X requires 3 hours of labour, 1 machine hour and 3 kg of raw material, and product Y requires 5 hours of labour, 1 machine hour, and 2 kg of raw material. The profit on X is £5 per unit and the profit on Y is £4 per unit. The manufacturer wishes to maximise his weekly profit on the production of the two products.
   Find the product mix that maximises profit.

16.2 A company manufactures two types of Easter Eggs, A and B, which sell at a profit of 40p and 50p respectively. The eggs are processed through three stages of production, and the average times spent at each stage are

	Blending	Cooking	Packing
A	1 min	5 min	3 min
B	2 min	4 min	1 min

During a production run, the blending equipment is available for a maximum of 12 machine hours, the cooking equipment for a maximum of 30 machine hours, and the packing equipment for a maximum of 15 machine hours.

   (a) Determine how many eggs of each type should be produced in order to maximise profit.
   (b) Comment on the utilisation of resources.

16.3 A manufacturer of two commodities has to use three separate materials in their manufacture. The relevant input data is:

Material	Product A	B	Material available
x	3	2	84 units
y	4	5	140 units
z	1	3	63 units

He makes £1 profit on a unit of A, and £2 on a unit of B. What product mix will maximise his profit?

16.4 A manufacturer of electrical equipment produces kettles and irons. The average cost of production is £12 for kettles and £8 for irons. In addition the firm faces transport costs of 40p per kettle and 60p per iron. To maintain its sales the firm has to spend 10p in advertising costs on each item sold. The retail price for kettles and irons are respectively £18 and £11.20. The firm's budget allows a maximum of £4,800 on production costs, £240 on transport costs and £500 on advertising costs. Find the output that maximises profit.

16.5 A firm producing three products has a bottleneck in production at a particular point of the productive process involving two machines. It realises there is something wrong with its product mix and wishes to feed into the bottleneck that combination which utilises the available machine time to its fullest extent and is also the most profitable. The relevant data showing the time spent on the production of each product is:

	Product x	y	z	Time available
Machine 1	24 min	60 min	72 min	240 h
Machine 2	60 min	72 min	48 min	240 h
Profit/unit	£1	£2	£2	

State the objective function and the constraints and deduce the optimum product mix. Check that capacity is fully utilised.

16.6 A firm makes three products A, B and C. Inputs per unit are

	Machine time	Labour time	Raw materials	Profit
a.	3 h	1 h	2 tons	£4
b.	2 h	2 h	1 ton	£6
c.	1 h	2 h	3 tons	£5
Available per week	218 h	134 h	150 tons	

He has a contract to supply 10 units of each product per week. Derive a solution matrix which will give maximum profit.

16.7 Let us consider an extension to the tailor problem. A second tailor is producing four products, the input requirement of each being

	Labour	Material	Lining
Coats	5 h	6 yd	4 yd
Jackets	4 h	4 yd	3 yd
Slacks	2 h	3 yd	1 yd
Waistcoats	2 h	1 yd	1 yd

His labour force is sufficient to provide 750 hours per week. A wholesaler can supply him with up to 950 yards of cloth and 560 yards of lining per week. Labour costs 50p per hour, cloth £1 per yard and lining 25p per yard.

After carefully assessing the prices charged by competitors, he decides to publish the following trade prices.

Coats   £10.10
Jackets   £ 7.25
Slacks   £ 4.65
Waistcoats   £ 2.50

If his objective is to maximise profit, what is the objective function? A department store buying manager sees potential in the tailor's designs, and a contract is arranged to supply 30 coats, 40 jackets, 50 pairs of slacks and 10 waistcoats per week. What are the constraints on the objective function?

Find a solution matrix to satisfy the objective.

16.8 A farmer has two types of pig meal which he has to combine to produce a satisfactory diet for his pigs.

A kilogram of Swillo contains 32 grams of nutrient A, 4 grams of nutrient B, and 4 grams of nutrient C.

A kilogram of Fillapig contains 16 grams of nutrient A, 8 grams of nutrient B, and 20 grams of nutrient C.

The pigs require at least 12 kilograms of pig meal each day, including a minimum of 256 grams of nutrient A, 56 grams of nutrient B and 80 grams of nutrient C.

   (a) Formulate a system of inequalities to represent the information above.
   (b) Find the least cost combination of pig meals compatible with the food requirements if a kilogram of Swillo cost £2 and a kilogram of Fillapig costs £3.
   (c) If the price of Fillapig rises to £10 per kilogram, the price of Swillo remaining constant, advise the farmer.

16.9 Suppose that ideally a batch of animal food should contain at least 270 units of vitamin A, 100 units of vitamin B and 190 units of vitamin C. The vitamin content is put into the food by adding certain additives. Additive X contains 2 units of each vitamin per gram and costs 5p per gram. Additive Y contains 3 units of vitamin A, and one unit each of vitamins B and C per gram and costs 6p. Additive Z contains 3 of vitamin A, 1 of B and 6 of C, costing 8p per gram. A manufacturer wishes to make a batch of this feed at a minimum cost; how many grams of X, Y and Z should be added?

17 Transportation and assignment

17.1 Introduction

When considering linear programming methods, we have so far confined our attention to production problems – maximum profit product – mix, or minimum cost combination of inputs, and blending problems. The time has now come to widen the horizons of linear programming techniques, and introduce various techniques which will simplify the method under certain conditions. Firstly, let us see how linear programming can help the transport manager to solve routing problems. Suppose a firm is producing a certain product in two separate factories. Factory A produces 75 units and factory B 25 units per week. The whole output is sold to three distributors, who take 45, 35 and 20 units per week. The transport manager has calculated the unit cost of moving the output to the distributors as follows:

	to distributor		
	a	b	c
from factory A	£3	£4	£6
B	£1	£4	£3

Thus it costs £3 to move a unit from factory B to distributor c, £4 to move a unit from Factory A to distributor b etc.

17.2 The Simplex form of the transportation problem

Let us suppose amounts X_{11}, X_{12} and X_{13} are distributed from factory A to distributors a, b and c respectively. Also that amounts X_{21}, X_{22} and X_{23} are distributed from factory B to each of the three distributors. The *transportation matrix* would look like this:

	to distributor		
	a	b	c
from factory A	X_{11}	X_{12}	X_{13}
B	X_{21}	X_{22}	X_{23}

Notice the convenience of using suffixes to X when stating a general solution to such problems: X_{13} means the quantity in the first row, third column. Thus the first digit of the suffix locates the row, and the second locates the column.

We can now state that the quantity distributed form Factory A is:

$X_{11} + X_{12} + X_{13}$

and as we wish the quantity distributed to be the same as the quantity produced:

$X_{11} + X_{12} + X_{13} = 75$

Likewise, for factory B:

$X_{21} + X_{22} + X_{23} = 25$

The quantity received by distributor a is:

$X_{11} + X_{21}$,

and as we wish distributor a to receive 45 units:

$X_{11} + X_{21} = 45$

Finally, from the restriction placed on distributors b and c:

$X_{12} + X_{22} = 35$

and $X_{13} + X_{23} = 20$

To find the cost of the allocation we note that for each unit sent from factory A to warehouse a, a cost of £3 is incurred, so the cost of sending X_{11} units is $3X_{11}$. Using similar arguments for the other quantities, the total allocation cost is:

$3X_{11} + 4X_{12} + 6X_{13} + X_{21} + 4X_{22} + 3X_{23}$

As the objective is to minimise cost, the problem becomes:

$$\text{Minimise } C = 3X_{11} + 4X_{12} + 6X_{13} + X_{21} + 4X_{22} + 3X_{23}$$

Subject to		
$X_{11} + X_{12} + X_{13} = 75$		(1)
$X_{21} + X_{22} + X_{23} = 25$		(2)
$X_{11} + X_{21} = 45$		(3)
$X_{12} + X_{22} = 35$		(4)
$X_{13} + X_{23} = 20$		(5)
All variables $\geqslant 0$		

17.3 A first feasible solution: the 'Northwest Corner' method

Of course, it would be perfectly possible to solve this problem using methods outlined in the previous chapter, but there is a better way. First, we combine the transportation matrix and the cost matrix like this:

3	4	6
X_{11}	X_{12}	X_{13}
1	4	3
X_{21}	X_{22}	X_{23}

The quantities X are the amounts allocated, and the amounts in the top right-hand corner of each cell represent the allocation costs. Now all linear programming problems that we have investigated involved finding a first feasible solution, and improving on it until an optimum is reached. This method (called the Transportation Method) is no exception. A first feasible solution can be obtained by the 'Northest Corner' method, so called because we consider the cell in the top left-hand corner first. We then make the amount allocated to this cell (X_{11}) as large as possible. Now we know that

$$X_{11} + X_{12} + X_{13} = 75 \tag{1}$$
and
$$X_{11} + X_{21} \quad\quad = 45 \tag{3}$$

As all variables must be non-negative, it follows that it is equation (3) that restricts the size of X_{11}, i.e $X_{11} \leqslant 45$. Now if X_{11} is to be as large as possible, then $X_{11} = 45$, $X_{21} = 0$, and the restrictions now become

$$X_{12} + X_{13} = 30 \tag{1}$$
$$X_{22} + X_{23} = 25 \tag{2}$$
$$X_{12} + X_{22} = 35 \tag{4}$$
$$X_{13} + X_{23} = 20 \tag{5}$$

We now turn to the second cell in the first row (X_{12}) and make this quantity as large as possible. From equation (1) we obtain that $X_{12} = 30$, and $X_{13} = 0$ (equation (4) does not restrict the size of X_{12}) and the restrictions now become

$$X_{22} + X_{23} = 25 \tag{2}$$
$$X_{22} \quad\quad = 5 \tag{4}$$
$$X_{23} = 20 \tag{5}$$

As the three restrictions are consistent with each other, we now have a feasible solution. The matrix now looks like this

x_{11} 3	x_{12} 4	x_{13} 6
45	30	0
x_{21} 1	x_{22} 4	x_{23} 3
0	5	20

The cost of this solution is $3 \times 45 + 4 \times 30 + 4 \times 5 + 3 \times 20 = £335$.

17.4 Finding an improved feasible solution

Returning to our earlier example, we know that the output can be allocated to the distributors at a cost of £335. Let us now see if it is possible to find another allocation at a lower cost. As the system now stands, factory A does not supply any units to distributor c, nor factory B to distributor a. In other words, all the cells are *occupied* except X_{13} and X_{21}. Now X_{21} is the cell with the lowest transport cost: it costs only £1 to supply distributor a from factory B. It would seem sensible then, to let factory B supply distributor a.

Let us investigate the effect on cost of supplying distributor a with one unit from factory B. How will this affect the solution? Cell X_{21} will now become occupied, having one unit. However, factory A will no longer supply distributor a with 45 units, as this would mean that distributor a would be receiving a total of 46 units – one more than he required. Hence cell X_{12} would contain 44 units. This now means that factory A now has a surplus unit, which we can distribute to cell X_{12} (notice that we do not put it in cell X_{13}, the other unoccupied cell – for the moment we are interested only in unoccupied cell X_{21}). If we remove one unit from cell X_{22}, the allocation would be feasible.

Thus we have

$X_{11} = 44$ (saving £3 in transportation costs)
$X_{21} = 1$ (costing £1 more in transportation costs)
$X_{12} = 31$ (costing £4 more in transportation costs)

and

$X_{22} = 4$ (saving £4 in transportation costs)

The net effect on cost of allocating a unit to cell X_{21} is

$$-3 + 1 + 4 - 4 = -£2$$

i.e such a move would save £2. We can conclude that every unit we could allocate to cell X_{21} would reduce transport costs by £2, and hence, it would pay us to move as much as possible to the cell.

How much can we allocate to cell X_{21}? For every unit that is allocated to X_{21}, we must remove a unit from both X_{11} and X_{22}. We can remove only 5 units from X_{22} – if we removed more than this we would have a negative (and hence non-feasible) solution. Hence we can move a maximum of 5 units into X_{21}. This would reduce transport costs by £10 (£2 per unit). The solution would now look like this:

x_{11} 3	x_{12} 4	x_{13} 6
40	35	0
x_{21} 1	x_{22} 4	x_{23} 3
5	0	20

You should satisfy yourself that this solution is feasible. The cost of this allocation is $3 \times 40 + 4 \times 35 + 1 \times 5 + 3 \times 20 = £325$. This agrees with the predicted cost reduction.

17.5 Reallocation routes

Let us look a little more closely at the problem of allocating into a chosen unoccupied cell, because it can be tricky. Suppose the following allocation had been selected.

X_{11}	X_{12}	X_{13}	X_{14}	X_{15}
77 −			38	+
X_{21}	X_{22}	X_{23}	X_{24}	X_{25}
9 +	35			51 −
X_{31}	X_{32}	X_{33}	X_{34}	X_{35}
	39	14		
X_{41}	X_{42}	X_{43}	X_{44}	X_{45}
		48		

Thus there are four factories supplying five distributors. Suppose it was decided that in future factory (1) should supply warehouse 5, i.e. cell X_{15} should become occupied − how could this be achieved? If distributor (5) is to be supplied from factory (1), he must receive less from factory (2). We mark cell X_{15} with a plus sign, and X_{25} with a minus. Now if factory (1) is to supply distributor (5) it must reduce its supply to other distributors − otherwise it will be asked to supply more than it can produce. It cannot supply less to distributor (4), as it is distributor (4)'s only supplier, and if we continue to use occupied cells only there would be no method of sympathetically increasing the amount allocated to distributor (4). Hence, it must supply less to distributor (1) and cell X_{11} is marked with a minus. Distributor (1) must now receive more from another factory, and as factory (2) now has surplus output cell X_{21} is marked with a plus. A reallocation route has now been obtained.

Using this route, we see that less is to be allocated to X_{11} and X_{25}. As negative allocations are not permitted, we can remove up to 51 units from both these cells. Thus 51 units is the most that can be allocated into X_{15}. Adding and subtracting 51 units as appropriate we obtain the new allocation. You can check the feasibility of the new solution: the row totals and column totals are the same as the initial allocation.

x_{11}	x_{12}	x_{13}	x_{14}	x_{15}
26			38	51
x_{21}	x_{22}	x_{23}	x_{24}	x_{25}
60	35			
x_{31}	x_{32}	x_{33}	x_{34}	x_{35}
	39	14		
x_{41}	x_{42}	x_{43}	x_{44}	x_{45}
		48		

Let us now state the rules for finding a reallocation route.

(1) Mark the unoccupied cell chosen for allocation with a plus.

(2) Examine the row containing the chosen cell, and mark the other occupied cell with a minus. If there are two occupied cells, mark the one which contains another cell within its column.

(3) Scan the column containing this cell, and mark the cell that contains another occupied cell within its row with a plus.

(4) Continue the process until each row that contains a plus also contains a minus (and vice versa) and until the same conditions apply to columns. A unique reallocation route will then have been found.

(5) The most that can be allocated into the chosen unoccupied cell is found by examining those cells containing minus signs. The smallest quantity in these cells is the most that can be moved.

Sometimes the reallocation route is quite involved. If we wanted to allocate into cell X_{45}, then the route would be:

x_{11}	x_{12}	x_{13}	x_{14}	x_{15}
26 +			38	51 −
x_{21}	x_{22}	x_{23}	x_{24}	x_{25}
60 −	35 +			
x_{31}	x_{32}	x_{33}	x_{34}	x_{35}
	39 −	14 +		
x_{41}	x_{42}	x_{43}	x_{44}	x_{45}
		48 −		+

A maximum of 39 units could be moved into cell X_{45}.

If we reallocate using occupied cells only (and we have seen that it is logical to do so as we are interested in one unoccupied cell only) then the Northwest Corner method gives a unique reallocation route. Sometimes it is not possible to use occupied cells only and we shall investigate this later. Finding the reallocation route is of cardinal importance in solving the transportation problem, and you should make sure you have mastered the technique before proceeding.

17.6 The optimum solution

We left our original problem looking like this:

x_{11} 3	x_{12} 4	x_{13} 6
40 −	35	 +
x_{21} 1	x_{22} 4	x_{23} 3
5 +		20 −

The 'Northwest Corner' solution was improved upon by allocating into cell X_{21} and clearing cell X_{22}. Can we get an even better solution? The only other solution possible would be to allocate into cell X_{13}. The route for this allocation is marked on the matrix, and we can see that the cost of moving one unit into cell X_{13} would be $+6+1-3-3 = +£1$. Thus, each unit moved into cell X_{13} would increase total cost by £1. Satisfy yourself that this is the case by moving the 5 units into X_{13}, and work out the new total cost. You will find that total cost has increased by £5. As we have now investigated all possible solutions we can see that the second solution was indeed optimal.

Let us now summarise the transportation method of linear programming. The transportation matrix is set up and the 'Northest Corner' technique used to obtain a first feasible solution. A low cost unoccupied cell is chosen and a route is found for allocating into that cell. The route is costed to see whether reallocation to that cell is worthwhile. The solution is optimal when no further reallocations are worthwhile.

17.7   Shadow costs

You are probably thinking that the method is tedious. It becomes most time consuming to examine the cost of all the alternative solutions, especially as the dimensions of the transportation matrix increase. Can you imagine trying to solve a ten factory, twenty distributor problem? We want a system that will identify immediately which of the alternative solutions will increase and which will decrease the cost of allocation.

To obtain such a system, we will return to the 'Northwest Corner' solution of our original problem:

X_{11} 3	X_{12} 4	X_{13} 6
45	30	
—	+	
X_{21} 1	X_{22} 4	X_{23} 3
	5	20
+	—	

We found that by allocating one unit to cell X_{21}, total cost would change by

$$-3+4+1-4 = -£2$$

i.e total cost would decrease by £2, so it is worthwhile moving as many units as possible into this cell. Now let us examine the change in total cost if we allocated one unit to cell X_{13}.

The change in total cost is

$$+6 - 3 + 4 - 4 = \text{£}3.$$

As total cost would rise by £3 per unit moved, it would not be worthwhile allocating to cell X_{13}.

We will now attempt to predict these changes in total cost using a method called *fictitious costing*. Such costs have no real meaning, but they do give an efficient method of highlighting those cells that would give a reduction in total cost. We start by supposing that the allocation cost of an *occupied* cell is made up from two components – the dispatch cost incurred by the factory and the reception cost incurred by the distributor. We will enter dispatch costs at the end of each row, and reception costs at the foot of each column.

Let us begin by examining the occupied cell X_{11} which has an allocation cost of £3 per unit. We know that Dispatch cost (for factory A) + Reception cost (for Distributor I) = £3. If we know the reception cost, then we could easily obtain dispatch cost by subtraction. We will suppose (quite arbitrarily) that Distributor I has a zero reception cost: this means that Factory A's dispatch cost is $3 - 0 = 3$.

	I		II		III		Dispatch costs ▼
A	x_{11}	3	x_{12}	4	x_{13}	6	3
	45		30				
B	x_{21}	1	x_{22}	4	x_{23}	3	
			5		20		
Reception costs ▶	0						

We can now calculate the remaining dispatch and reception cost. Using cell X_{12}, reception cost for distributor II = $4 - 3 = 1$.

	I		II		III		Dispatch costs ▼
A	x_{11}	3	x_{12}	4	x_{13}	6	3
	45		30				
B	x_{21}	1	x_{22}	4	x_{23}	3	
			5		20		
Reception costs ▶	0		1				

The next cell to use is X_{22} (we cannot use X_{13} or X_{21} as they are both empty).
Dispatch cost for factory B = $4 - 1 = 3$.

	I		II		III		Dispatch costs ▼
A	x_{11}	3	x_{12}	4	x_{13}	6	3
	45		30				
B	x_{21}	1	x_{22}	4	x_{23}	3	3
			5		20		
Reception costs ▶	0		1				

Finally, using cell X_{23}, reception cost for distributor III $= 3 - 3 = 0$.

Dispatch costs ▼

	I	II	III	
A	x_{11} 3 45	x_{12} 4 30	x_{13} 6	3
B	x_{21} 1	x_{22} 4 5	x_{23} 3 20	3
Reception costs ►	0	1	0	

Having found all dispatch and reception costs, the next stage is to calculate *shadow* costs. For *empty* cells,

Shadow cost = dispatch cost + reception cost

For cell X_{21}, shadow cost $= 3 + 0 = 3$
For cell X_{13}, shadow cost $= 3 + 0 = 3$

The saving in total cost by allocating one unit into each empty cell can now be calculated by subtracting allocation cost from shadow cost. So the saving by allocating one unit to X_{21} is $3 - 1 = £2$, and the saving by allocating one unit to X_{13} is $3 - 6 = -£3$ (i.e. cost would increase). The cost change is entered in the bottom right hand corner of each empty cell.

Dispatch costs ▼

	I	II	III	
A	x_{11} 3 45	x_{12} 4 30	x_{13} 6 −3	3
B	x_{21} 1 2	x_{22} 4 5	x_{23} 3 20	3
Reception costs ►	0	1	0	

What can we conclude? The cost of not using cell X_{21} is £2, so we would reduce total cost by £2 for every unit that we allocated to that cell. Likewise, the cost of not using cell X_{13} is $-£3$, so we would increase cost by £3 for every unit that we allocated to X_{13}. (Notice that this agrees precisely with our conclusions at the start of this section.) *Thus we will allocate to cells where the cost of not using them is positive, and avoid those cells where the cost of not using them is negative. Also we will concern ourselves first with the cell that has the greatest positive cost of remaining unused.* The shadow cost method has the great advantage of finding such costs without the necessity of identifying all the reallocation routes first. In fact we find only one allocation route: the one for the unoccupied cells that reduces cost most. Let us now examine a matrix with greater dimensions.

x_{11} 4	x_{12} 5	x_{13} 5	x_{14} 1	x_{15} 2	
90	36				4
	−		+ 1	1	
x_{21} 4	x_{22} 4	x_{23} 2	x_{24} 1	x_{25} 2	
	22	46			3
	+	−	0		
x_{31} 5	x_{32} 6	x_{33} 3	x_{34} 2	x_{35} 3	
		27	38		4
		+	−		
x_{41} 5	x_{42} 1	x_{43} 5	x_{44} 4	x_{45} 2	
13				22	5
	5				
0	1	−1	−2	−3	

Reception costs and dispatch costs have been calculated in the usual way, only the positive costs of not using the unoccupied cells have been entered, so that we can see at a glance the cells we should concentrate on. It is easy to decide whether such costs will be positive − if shadow cost exceeds allocation costs, then such costs will indeed be positive. The matrix shows that by allocating into either X_{14} or X_{42} we could reduce costs. It also tells us to concentrate first on X_{42}, as this gives the greatest saving per unit allocated. Incidentally, the shadow cost method gives the correct cost of not allocating to an empty cell even when the route is not straightforward. The route for reallocating into X_{14} is marked on the matrix, and the change in cost per unit moved is $+1 - 2 + 3 - 2 \div 4 - 5 = -£1$. Hence the cost of not using this route is £1, which agrees with the result obtained by shadow costs.

This enlarged example clearly illustrates the labour saved using the shadow cost method. The method outlined earlier would have involved finding the 12 other routes, and calculating the cost of using each!

17.8 Degeneracy

Earlier it was stated that a unique reallocation route for each unoccupied cell could be found. In fact, this is true only if the number of occupied cells is $n + m - 1$, where n is the number of rows, and m is the number of columns. If the number of occupied cells is less than this, then it may not be possible to find a reallocation route using occupied cells only. Moreover, it will be impossible to find all the shadow costs. Such matrices are called *degenerate*, and we require a method for dealing with them.

The advantage of the 'Northwest Corner' method is that it avoids degeneracy. Why then, you may ask, should we bother to investigate a method for dealing with it? Well the 'Northwest Corner' method is not a very efficient method of obtaining a first feasible solution – we would be nearer the optimal solution if we allocated to the low allocation cost cells first. But this system of allocation often results in a degenerate matrix. Secondly, it sometimes happens that a non-degenerate matrix goes degenerate after reallocation. Consider the following matrix.

x_{11}	x_{12}
20	30
x_{21}	x_{22}
	20

In this case, $n + m - 1 = 3$, the same as the number of occupied cells. Now suppose it was found to be worthwhile to allocate into cell X_{21}, the resultant matrix would be

x_{11}	x_{12}
	50
x_{21}	x_{22}
20	

which is degenerate.

Let us suppose we were presented with the following feasible solution to a transportation problem. We want to know whether there is a better solution.

x_{11}	1	x_{12}	2	x_{13}	3	
15				30		1
x_{21}	2	x_{22}	5	x_{23}	4	
		35				
x_{31}	1	x_{32}	3	x_{33}	2	
20						1
0				2		

We begin as usual by assigning a zero reception cost to column 1. This gives despatch costs of 1 to rows 1 and 3. Using cell X_{13} we can obtain a reception cost of 2 for column 3. It is not possible to calculate any other costs as there are insufficient occupied cells. The way out of this problem is to treat one of the unoccupied cells as if it were occupied – which one should we choose? Obviously, we would not choose X_{33} as we already have the dispatch and reception costs relating to that cell. We would choose a cell for which we have either a dispatch cost or a reception cost but not both, and generally, it is better to choose a cell with a low allocation cost. We will treat X_{21} as though it were occupied by allocating to it a small quantity epsilon (ε). Epsilon is so small that it does not affect the feasibility of the solution, nor does it affect the total cost. We can now complete the calculation of dispatch and reception costs:

X_{11} 1	X_{12} 2	X_{13} 3	
15	ε +	30 —	1
X_{21} 2	X_{22} 5	X_{23} 4	
  2	35 —	 + 2	4
X_{31} 1	X_{32} 3	X_{33} 2	
20		 1	1
0	1	2	

Allocating 30 units to X_{23} gives:

X_{11} 1	X_{12} 2	X_{13} 3	
15 —	$\varepsilon + 30$ +		1
X_{21} 2	X_{22} 5	X_{23} 4	
  + 2	5 —	30	4
X_{31} 1	X_{32} 3	X_{33} 2	
20			1
0	1	0	

Now allocating 5 units to X_{21}:

x_{11} 1 10	x_{12} 2 $\varepsilon+35$	x_{13} 3	1
x_{21} 2 5 +	x_{22} 5	x_{23} 4 30 −	2
x_{31} 1 20 −	x_{32} 3	x_{33} 2 + 1	1
0	1	2	

Finally, allocating 20 units to X_{33} gives:

x_{11} 1 10 −	x_{12} 2 $\varepsilon+35$	x_{13} 3 + 0	1
x_{21} 2 25 +	x_{22} 5	x_{23} 4 10 −	2
x_{31} 1	x_{32} 3	x_{33} 2 20	0
0	1	2	

This is an optimal solution, though there is a second solution obtained by allocating to X_{13} (the cost of not using X_{13} is zero). Notice that the second optimal solution is degenerate.

x_{11} 1	x_{12} 2	x_{13} 3	
ε	$\varepsilon+35$	10	1
x_{21} 2	x_{22} 5	x_{23} 4	
35		0	2
x_{31} 1	x_{32} 3	x_{33} 2	
		20	0
0	1	2	

As an optimal solution can be degenerate, this gives a third reason for mastering the technique of handling degeneracy. In the solutions above, we can drop the epsilon as it is so insignificantly small that it affects neither feasibility nor cost. In practice epsilon is dropped when the matrix ceases to be degenerate.

17.9   A better method of obtaining a first feasible solution

We noted earlier that we can improve on the Northwest Corner method of finding a first feasible solution. This method involves finding the low cost cells and allocating to them first. This will give a solution nearer to the optimum. Let us examine the following example to see how the method works.

Output of four factories: 110, 80, 60 and 50 units.
Demand of five distributors: 140, 70, 40, 30 and 20 units.
Allocation costs:

	1	2	3	4	5
1	£4	£5	£5	£6	£1
2	£2	£4	£4	£3	£5
3	£3	£2	£4	£5	£5
4	£3	£5	£2	£3	£4

The transportation matrix is presented in the usual form, except for a new first row which gives distributor's demand, and a new first column which gives factory output.

	140̶ 60	70̶ 10	40̶	30̶ 20	20̶	
110̶ 90̶ 30̶ 20	x_{11} 4 60 +	x_{12} 5 10	x_{13} 5	x_{14} 6 20 −	x_{15} 1 20	4
80̶	x_{21} 2 80 −	x_{22} 4	x_{23} 4	x_{24} 3 + 1	x_{25} 5	2
60̶	x_{31} 3	x_{32} 2 60	x_{33} 4	x_{34} 5	x_{35} 5	1
50̶ 10̶	x_{41} 3	x_{42} 5	x_{43} 2 40	x_{44} 3 10	x_{45} 4	1
	0	1	1	2	−3	

The cell with the lowest allocation cost is X_{15}, and we begin by allocating as much as possible to that cell. Now although factory (1) can supply 110 units, distributor (5) can only take 20 units, hence 20 is the most we can allocate to X_{15}. Now we will allow the first row to indicate how much is to be allocated to each distributor, and the column to indicate the amount of output undistributed. Distributor (5) has his full quota, and factory (1) has $110 - 20 = 90$ units undistributed. We show this information at the beginning of column 5 and row 1. We now examine cells with £2 allocation costs (X_{21}, X_{32}, and X_{43}). We allocate as much as possible to these cells, and adjust the row and column headings accordingly. Now we examine cells with £3 allocation costs. We cannot allocate to X_{24} (factory 2's output has all been allocated) nor X_{31} (factory 3's output has all been allocated). We can allocate to either X_{41} or X_{44}. If we do not allocate to X_{41}, then we must allocate to some other cell in row 1. Likewise, if we do not allocate to X_{44} we must allocate to some other cell in column (4). Now as the cells in column (4) have higher allocation costs than column (1), it would seem more logical to allocate to X_{44}. The allocation is now fixed and we have no further choices. The remaining 90 units produced by factory 1 must be allocated as shown. Shadow costs are calculated in the unusual fashion, and the only way total cost can be reduced is to allocate into X_{24}. The reallocation route is marked, and allocating 20 units into X_{24} gives:

x_{11} 4	x_{12} 5	x_{13} 5	x_{14} 6	x_{15} 1	
80	10			20	4
x_{21} 2	x_{22} 4	x_{23} 4	x_{24} 3	x_{25} 5	
60			20		2
x_{31} 3	x_{32} 2	x_{33} 4	x_{34} 5	x_{35} 5	
	60				1
x_{41} 3	x_{42} 5	x_{43} 2	x_{44} 3	x_{45} 4	
		40	10		2
0	1	0	1	—3	

This is the optimum allocation

17.10 Solution of transportation matrices when supply and demand are not equal

So far, we have considered examples where supply and demand are matched exactly. This situation is seldom found in practice – shortages of supply or excess capacity are most common. Consider the case where a good is a by-product of some other good – it would be almost impossible to match supply to demand. Let us see how we can deal with such situations.

Output   65,   25,   10
Demand   45,   25,   20

Thus, supply exceeds demand by 10 units. The allocation costs are:

    £2   £3   £4
    £1   £4   £2
    £5   £1   £3

We deal with this problem by supposing that the surplus output is put into store, and that the cost of moving the output into store is zero. We introduce an extra column to represent the store and assign to it zero allocation costs. The amount to be allocated to this column is ten units. Allocating according to the 'least cost cell' first method, and calculating shadow costs in the usual fashion we have:

	4̶5̶ 2̶0̶	2̶5̶ 1̶5̶	2̶0̶	1̶0̶	
6̶5̶ 4̶5̶ 3̶0̶ 1̶0̶	x_{11} 2 20 +	x_{12} 3 15	x_{13} 4 20 −	x_{14} 0 10	2
2̶5̶	x_{21} 1 25 −	x_{22} 4	x_{23} 2 + 1	x_{24} 0 2	1
1̶0̶	x_{31} 5	x_{32} 1 10	x_{33} 3 .	x_{34} 0	0
	0	1	2	−2	

Suppose we had decided not to store the surplus output, but to destroy it. We could then have assigned a very high cost X to each cell of a 'disposal' column – i.e we would have assumed the high allocation costs represent the loss of revenue. Would this have made any difference to our solution? Try it for yourself – what can you conclude?

If demand exceeds supply, then we use exactly the same method, but this time introducing a false row.

Allocating into cell X_{23} gives

x_{11} 2 40	x_{12} 3 15	x_{13} 4	x_{14} 0 10	2
x_{21} 1 5	x_{22} 4	x_{23} 2 20	x_{24} 0	1
x_{31} 5	x_{32} 1 10	x_{33} 3	x_{34} 0	0
0	1	1	−2	

– the optimum solution.

17.11 A problem involving maximisation

Suppose that rather than calculating the cost of allocating the output of a given number of factories to a given number of distributors, a transport manager calculates what would be the profit earned on all the possible allocations. For example, suppose the output of three factories could be sold to four distributors at the following rates of profit:

£2	£3	£4	£6
£1	£4	£3	£2
£5	£1	£3	£2

The output of the factories are 65, 25 and 10 units, and the demand of the distributors are 45, 25, 20 and 10 units. The problem is to allocate the output so as the maximise the profit.

Think back to the Simplex method – it was used to find solutions to objective functions which are to be maximised. If we wished to minimise, we solved the dual problem. Now the transportation method is completely the reverse; if we wish to minimise we solve the primal problem, and if we wish to maximise we solve it dual. Can you remember that we find the dual by turning the primal problem 'inside out'? How can we turn the primal transportation problem inside out? We select the greatest unit profit (in this case it is £6) and subtract all other profits from this.

4	3	2	0
5	2	3	4
1	5	3	4

Compare the two matrices carefully – the component which had the greatest value in the profit matrix has the least value in the transformed matrix. Also, the component which had the least value in the profit matrix has the greatest value in the transformed matrix. In fact, the order of magnitude of the components has been completely reversed. Now if we minimise the transportation matrix using the transformed components, surely we will be finding that allocation which *maximized* profit. Setting up this matrix in the usual fashion we have:

	45̶ 35	25̶	20̶	10̶	
65̶ 55̶ 35̶	x_{11} 4 35	x_{12} 3	x_{13} 2 20	x_{14} 0 10	4
25̶	x_{21} 5 ε	x_{22} 2 25	x_{23} 3	x_{24} 4	5
10̶	x_{31} 1 10	x_{32} 5	x_{33} 3	x_{34} 4	1
	0	−3	−2	−4	

This is the optimum allocation, and to find the total profit earned by this allocation, we must use the actual profit figures, not the transformed components, i.e. $2 \times 35 + 4 \times 20 + 6 \times 10$ etc.

17.12 An assignment problem

A recruitment officer has five different vacancies to fill, and five men available to fill them. He devises an aptitude test for each vacancy, and administers each test to the applicants. Assessment is on a ten point scale, and a low mark indicates a high aptitude. Thus a score of one would mean the candidate is highly suitable, and a score of ten would indicate that the candidate was most unsuitable. The candidates were assessed as follows:

		Vacancy				
		I	II	III	IV	V
	A	8	7	8	4	1
	B	3	6	2	3	4
Man	C	7	2	1	2	3
	D	4	6	5	5	2
	E	8	8	5	7	1

The objective of the recruitment officer is to allocate men to the tasks in such a fashion that their effectiveness is maximised. It would be possible to solve this problem using the transportation method. If we define X_{ij} as meaning that man i does job j, then clearly either $X_{ij} = 1$ (he does the job) or $X_{ij} = 0$ (someone else does the job). A feasible solution might look like this:

I	II	III	IV	V
8	7	8	4	1 **1**
3	6	2 **1**	3	4
7 **1**	2	1	2	3
4	6 **1**	5	5	2
8	8	5 **1**	7	1

The problem is – how will we set this allocation to ensure it is optimal? We could use the method of the last chapter and find dispatch and reception costs; we would need $5 + 5 - 1 = 9$ occupied cells so we would have to allocate epsilon to 4 empty cells. Fortunately, there is a better method of solving such problems.

17.13 The Hungarian method

	Vacancy					row
	I	II	III	IV	V	minima
A	8	7	8	4	1	1
B	3	6	2	3	4	2
Man C	7	2	1	2	3	1
D	4	6	5	5	2	2
E	8	8	5	7	1	1

We can deduce from the matrix that man A is best employed at job V, man B is best employed at job III, etc. The row minima inserted at the end of the matrix gives the best rating obtained by each man. Suppose we subtract the row minima from all the scores in the corresponding row (i.e. subtract 1 from row 1, 2 from row 2, 1 from row 3, etc), then the resulting zeros would highlight the best job for each man.

	I	II	III	IV	V
A	7	6	7	3	0
B	1	4	0	1	2
Man C	6	1	0	1	2
D	2	4	3	3	0
E	7	7	4	6	0

A glance at this matrix confirms that man A, man D and man E are all best suited to job V while man B and man C are best suited to job III. Now let us examine this matrix from the point of view of the vacancies. The column minima gives the best rating obtained at each job, i.e.

1   1   0   1   0

and if we subtract the column minima from all the elements in the corresponding columns, then the resulting zeros would again highlight areas of greatest effectiveness.

	Vacancy				
	I	II	III	IV	V
A	6	5	7	2	0
B	0	3	0	0	2
Man C	5	0	0	0	2
D	1	3	3	2	0
E	6	6	4	5	0

We will now examine this matrix to see whether the position of the zeros would give a feasible allocation. To do this, we will copy out the matrix again, this time inserting the zeros only.

	Vacancy				
	I	II	III	IV	V
A					0
B	0		0		0
Man C		0	0	0	
D					0
E					0

Column I contains only one zero, and so we will allocate man B to job I. Now row B indicates that man B would also be efficiently employed at jobs III and IV but as we have already allocated man B to job I, he cannot also do job III and job IV. Hence, we delete the other zeros in row B. Column II also contains one zero, so we allocate man C to job II, and remove the other two zeros from row C. Row A contains just one zero, so we allocate man A to job V and remove the other two zeros in column V. Unfortunately, we have now used all the zeros either to allocate (indicated by $\emptyset$ or we have eliminated then (indicated by $\emptyset$). However, we have not allocated men to jobs III and IV, so we have not reached an optimal solution. The solution suggested by the zeros is not feasible, not because it contains insufficient zeros, but because the zeros are not distributed in such a fashion to generate a feasible solution. We require a situation where we can allocate to each row and each column at least one unique zero, and the Hungarian method will test whether this is so. It involves drawing the minimum number of straight lines (either horizontally or vertically) so as to cover all the zeros in the matrix. let us see how this is done. We scan the matrix and find the row or column that contains the most zeros. Column V, row B and row C each contain 3 zeros and we can cover all the zeros with one vertical and two horizontal lines. All of the zeros have now been covered, and it is impossible to cover them all with less than three straight lines. Now this is a five row by five column matrix, and the Hungarian method states that if the minimum number of lines necessary to cover all the zeros is five, then we will have reached a feasible solution.

		Vacancy			
	I	II	III	IV	V
A	6	5	7	2	0
B	0	3	0	0	2
Man C	5	0	0	0	2
D	1	3	3	2	0
E	6	6	4	5	0

As we can cover all the zeros with three straight lines, we must introduce another zero into the matrix. To do this, we first locate the smallest uncovered element in the matrix (which in this case is one) and subtract this from all the uncovered elements. By selecting the smallest element, we have introduced the new zero in the most effective place. This new zero is our entering variable, and as we discovered when looking at the Simplex method, an entering variable requires a sympathetic departing variable. We achieve this by adding the smallest uncovered element to any element that lies on the intersection of lines covering the zeros. Thus, we add one to the element in row B and column V, and we add one to the element in row C and column V. (Of course, if we do not have a zero at one of the intersections, then we will not have a departing variable.) The matrix now looks like this:

		Vacancy			
	I	II	III	IV	V
A	5	4	6	1	0
B	0	3	0	0	3
Man C	5	0	0	0	3
D	0	2	2	1	0
E	5	5	3	4	0

We can never cover all the zeros with four lines, so we have not yet reached an optimal solution. We need to introduce another zero into the matrix and again look for the smallest uncovered element (one). Subtracting one from all the uncovered elements and adding one to the elements at the intersections of the lines gives the next new matrix.

		Vacancy			
	I	II	III	IV	V
A	5	3	5	0	0
B	1	3	0	0	4
Man C	6	0	0	0	4
D	0	1	1	0	0
E	5	4	2	3	0

It is impossible to cover all the zeros in the matrix with less than five straight lines, so the position of the zeros indicates the optimum solution.

		I	II	III	IV	V
	A				0	0
	B			0		0
Man	C		0	0	0	
	D	0			0	0
	E					0

Column (I) contains only one zero, so man D is allocated to job I. Now row D indicates that man D could also fill jobs IV and V, but he cannot do this as we have already allocated him to job 1. Hence, we delete the other two zeros in row D as they cannot form part of a feasible solution. Column (II) also contains one zero, so we allocate man C to job II, and eliminate the other two zeros in row C. Now column (III) has one zero, so we allocate man B to task III, and delete the other zero in row B. Column (IV) contains one zero so we allocate man A to job IV and delete the other zero in row A. Allocating man E to job V completes the solution. The men have now been allocated to the jobs in the most effective manner.

17.14 Summary of the Hungarian method

Let us now summarise the Hungarian method with a series of simple rules that you can apply to obtain the optimum allocation.

Step 1. Find the smallest element in each row and subtract it from all the elements in that row.

Step 2. Find the smallest element in each column and subtract it from all the elements in that column.

Step 3. Draw the minimum number of horizontal and vertical lines so as to cover all of the zeros. If the number of lines you have drawn is the same as the dimension of the matrix then go to step 7.

Step 4. Find the smallest uncovered element and subtract this number from all the uncovered elements.

Step 5. Add this number to any element at the intersection of two lines.

Step 6. Go to step 3.

Step 7. By examining the position of the zeros, extract from the matrix the optimum allocation.

Perhaps the greatest drawback of the Hungarian method is that it relies on human judgement to decide whether or not the minimum number of lines have been used to cover all the zeros. Sometimes we may think that we have used the minimum number of lines when in fact we could have used less. However, you can rest assured that if you have not used the minimum number of lines, then you will not be able to make an allocation. Consider the matrix below where it appears that five lines are necessary to cover all of the zeros.

	I	II	III	IV	V
A	5	3	5	2	0
B	1	3	0	0	4
Man C	6	0	0	0	4
D	0	4	2	0	0
E	6	2	3	4	0

When we attempt an allocation, we find that although we can allocate men to jobs I, II, and III, we have no suggestion to make for job IV and two men available for job V!

	I	II	III	IV	V
A					0
B			0	0	
Man C		0	0	0	
D		0		0	0
E					0

The matrix does not generate a solution because we could have covered all the zeros with four lines like this:

	I	II	III	IV	V
A	5	3	5	2	0
B	1	3	0	0	4
Man C	6	0	0	0	4
D	0	4	2	0	0
E	6	2	3	4	0

Before we leave the Hungarian method for assignment, it should be noted that we have used the method for minimising, but we could also use it for maximising. To do this, we would find the dual of the problem in the same way as we did when discussing transportation. We would find the largest element in the matrix and subject all other elements in the matrix from this. So to maximise this:

8	7	8	4	1
3	6	2	3	4
7	2	1	2	3
4	6	5	5	2
8	8	5	7	1

We would minimise this:

0	1	0	4	7
5	2	6	5	4
1	6	7	6	5
4	2	3	3	6
0	0	3	1	7

Review questions

17.1 The weekly output of three factories is:

Factory	A	B	C
	139	74	32

The weekly demand from 5 distributors is:

Distributor	a	b	c	d	e
Demand	75	65	55	40	10

and the unit allocation costs are

	a	b	c	d	e
A	£2	£3	£5	£6	£7
B	£3	£6	£5	£2	£1
C	£1	£7	£5	£3	£2

Allocate the output of the factories to the distributors at minimum cost.

17.2 A firm produces a certain product at three factories, and sells the output to four distributors. Output and demand are

Factory	A	B	C	
Output (weekly)	65	25	10	
Distributor	a	b	c	d
Demand (weekly)	45	25	20	10

The costs of distribution are

	a	c	d	b
A	£2	£4	£6	£3
B	£1	£3	£2	£4
C	£5	£3	£2	£1

If the objective is to minimise cost, form the linear programming model of the problem. Find the least cost of allocation.

17.3 Suppose an oil company has its refineries sited at Southampton, the Thames estuary, Merseyside and Milford Haven. One of the by-products of refining oil is phenol, which is used by five divisions of the company: Plastics Division of Birmingham, Paints Division of Manchester, Fibres Division of Leeds, Adhesives Division at Bristol, and Drugs Division at Nottingham. The company has its own fleet of vehicles, and it has been calculated that the cost of transporting phenol is 1p per unit per mile. Other relevant information is as follows

Phenol Production	
Southampton	115,000 units
Thames Estuary	95,000 units
Merseyside	53,000 units
Milford Haven	48,000 units
	311,000 units

Phenol Consumption	
Plastics Division	86,000 units
Paints Division	74,000 units
Fibres Division	62,000 units
Adhesives Division	38,000 units
Drugs Division	51,000 units
	311,000 units

Road Distances

	Birmingham	Manchester	Leeds	Bristol	Nottingham
Southampton	128	206	224	75	158
Thames Estuary	110	184	190	116	122
Merseyside	90	35	73	160	97
Milford Haven	167	208	248	146	216

The problem is to allocate phenol to the divisions at minimum cost.

17.4 The Thames refinery is put out of action by a strike, and it is decided to buy phenol from abroad. Assuming it costs more to import phenol than to produce it, find the optimum allocation.

17.5 Find the optimum allocation assuming that Drugs Division is closed, but the Thames refinery strike is over.

17.6 Test whether the following degenerate matrix is optimal (objective is least cost allocation):

	30	20	10	5
35	1 30	1 5	3	4
25	6	3 15	4 10	2
5	3	2	4	5 5

17.7 Assume that the costs in question 17.2 are in fact profits. Find the optimal allocation.

17.8 A vehicle manufacturer has divided production in the body shop to 45 per cent saloons, 45 per cent vans and 10 per cent convertibles. Each vehicle is capable of taking three different engines – 1000 cc., 1250 cc. and 1500 cc. The sales manager supplies suitable selling prices, and the accountant calculates the profit per vehicle to be as follows

	1000 cc.	1250 cc.	1500 cc.
Saloon	£100	£115	£130
Van	£160	£170	£180
Convertible	£60	£75	£90

If 70 per cent of engines produced are 1000 cc., 10 per cent are 1250 cc. and 20 per cent are 1500 cc., decide which engine should be fitted into which vehicle, and find the proportion of output for each variant. What is the total profit per 100 vehicles sold?

17.9 A property developer has four projects which it puts out to tender to four contractors. The contract bids are shown in the table below (all figures are million pounds).

		Project		
Contractor	I	II	III	IV
A	22	38	33	19
B	26	36	42	14
C	24	42	40	20
D	38	41	37	18

Assume that the developer considers it politic to award one contract to each bidder. Assign the projects to the contractors.

17.10 Suppose that the data in question 17.9 represents profits to the developer rather than bids. Assign the projects to the contractors such that the developer's profit is maximised.

17.11 A national plant-hire firm receives orders for cranes from firms in Preston, Taunton, Northampton, Oxford and York. It has a crane available in Southampton, Bristol, Birmingham, Nottingham and

Shrewsbury. The road distances are:

	Preston	Taunton	Northampton	Oxford	York
Southampton	233	87	106	65	236
Bristol	184	43	102	69	214
Birmingham	105	130	50	64	127
Nottingham	100	180	57	94	78
Shrewsbury	82	146	93	104	130

Decide on the allocation of cranes to meet the orders

17.12 If we use the Hungarian method, then we will have an optimum allocation for an $N \times N$ matrix when N lines are required to cover all the zeros. Let us return to the example worked in Section 17.13 and suppose that it was decided to scrap the last two vacancies. However, five candidates are still available. We now have a 5×3 matrix, but to use the Hungarian method we need a square matrix. We can still solve the problem by introducing two imaginary vacancies and assume that each candidate has a zero rating for these vacancies. Notice that there is no point in subtracting the row minima (they are all zero), so proceed directly to the column minima and find the optimal allocation.

17.13 It is decided to train six astronauts for a flight to Mars. They all receive a general training and then receive aptitude tests for their roles on the mission. The marks awarded were:

			Astronaut			
Test for	A	B	C	D	E	F
Flight Commander	21	5	21	15	15	28
Pilot	30	11	16	8	16	4
Navigator	28	2	11	16	25	25
Engineer	19	16	17	15	19	3
Back-up man	26	21	22	28	29	24
Communications	3	21	21	11	26	26

Suggest the crew composition.

17.14 Suppose it was decided to dispense with the back-up man. What should be the composition of the crew?

18 Network analysis: project planning

18.1 Introduction

We all know that if a project is to be undertaken efficiently, it must be planned, especially if a number of people are engaged on it. All too often, people work in isolation, quite ignorant of what other people working on the same project are doing. It is the task of management to co-ordinate efficiently the efforts of such people. In the past, planning has not been so important as it is today as projects were less complex – the rule of thumb method worked quite well. However, as projects have become more complex, management researchers have turned their attention increasingly to systematic planning. Consider the complexity of producing the prototype Concorde or an Apollo moonshot, and you will soon recognise the need for systematic planning. Shortly after the Second World War, researchers evolved a method called *network analysis*. The impact of this method has been quite dramatic, largely because it is applicable to such a wide variety of projects. It has been used to plan production projects, servicing projects, research projects, sales projects and military projects.

18.2 Network symbols

How does the method work? Well, the first stage of analysis is to divide the project into a number of different *activities*. An activity is merely a particular piece of work identifiable as an entity within the project. If, for example, the project under consideration is the servicing of a motor car, then one of the activities would be 'check the brakes for wear'. Now an activity within a network is represented by an arrow, with the description of activity written on it, as in Fig. 18.1

In addition to activities, we must also identify *events*. Events mark the point in time when an activity is complete and the next activity can be started. Events are represented by circles – see Fig. 18.2. The event ⓘ represents the point in time when the car is ready to have its brakes tested.

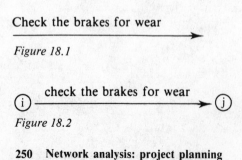

Check the brakes for wear

Figure 18.1

Figure 18.2

So far, we have not attempted to define a network and this we must now do. A network is a convenient method of showing the logical sequence of activities in a project. Suppose that in a certain project there are two activities A and B, and activity B cannot be started until activity A is completed. Using the network symbols, these activities can be represented as in Fig. 18.3.

Figure 18.3

The event $\widehat{x}$ represents the point of time when activity A is completed, but it also represents the point of time when activity B can begin. We can conclude that Fig. 18.3 is a true representation of the situation when activity B depends upon activity A. Now of course it is quite likely that two or more activities are dependent upon the same activity. The situation when neither activity D nor E can start until activity C is complete would be represented as in Fig. 18.4.

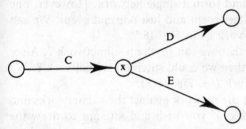

Figure 18.4

The event $\widehat{x}$ represents the point of time when activity C is completed, and also the point of time when activities D and E can start, so Fig. 18.4 clearly shows that D and E depend upon C. Also, it is likely that an activity depends upon more than one other activity. If activity H cannot start until activities G and F are both complete, then we would represent the situation as in Fig. 18.5.

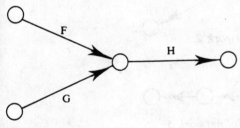

Figure 18.5

18.3 Precedence tables

You will probably have realised by now that the first task of network analysis is to sort out the logical sequence of activities, and we can summarise the logical sequence in a so-called *precedence table*. In this table we list all the activities and their immediately preceding activities.

Activity	Preceding activity
B, C and D	A
E and F	B
G	E
H	F
J	C
K	D
I	G and H
L	I, J and K

Constructing a precedence table is often the most difficult part of project analysis. Can you imagine how difficult it would be to construct a precedence table if the project involved the construction of a Polaris submarine? Obviously, the construction of a precedence table is, in essence, a team activity: all the experts involved in the project must be consulted.

Once we have compiled the precedence table we can begin to draw the network, and the first stage is to represent the relationships by a series of subnetworks (see Fig. 18.6).

We can now attempt to join the subnetworks together and form a single network. However, one important rule must be observed: there must be just *one* start event and just *one* end event. We can stitch together the first six subnetworks into a single network as in Fig. 18.7.

If we make activities G and H have the same end event, then we can stitch on subnetwork 7. Also, if we make activities I, J and K have the same end event, then we could stitch on subnetwork 8. We will then have represented the project with a single network (see Fig. 18.8).

You should carefully check the sequence of activities in the network against the subnetworks and precedence table. Of course, you can omit the subnetworks if you wish and attempt to draw the complete network directly from the precedence table. Whichever way you do it, you will seldom draw the network correctly the first time you try.

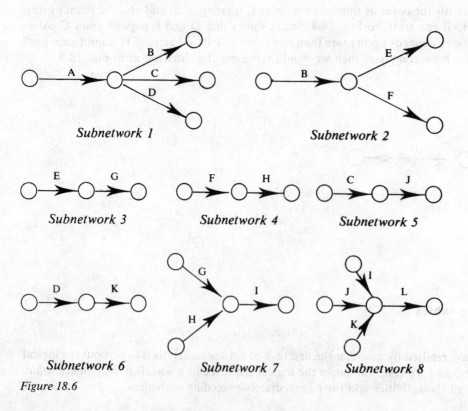

Subnetwork 1

Subnetwork 2

Subnetwork 3

Subnetwork 4

Subnetwork 5

Subnetwork 6

Subnetwork 7

Subnetwork 8

Figure 18.6

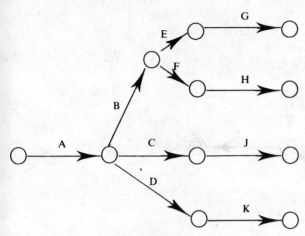

Figure 18.7

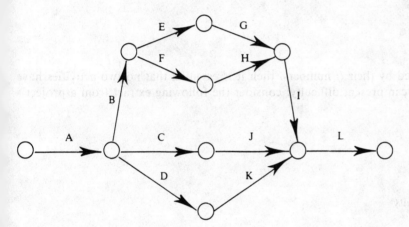

Figure 18.8

18.4 The *ij* event numbering rule

If the network above consisted of real rather than imaginary activities, then a description of each activity would be written above each arrow. It is convenient to use a coding system to describe a particular activity, and we do this by numbering the events according to the *ij* rule. The rule states that the event at the end of the activity must be assigned a greater number than the event at the beginning of an activity.

There is no single way of numbering the events, for the *ij* rules allows considerable latitude. One numbering system that obeys the rule is shown in Fig. 18.9.

But of course, there are many others. We can now describe activity A (remember that in fact it may have quite a lengthy description) by its *ij* numbers 1–2.

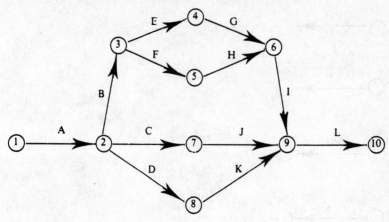

Figure 18.9

18.5 Dummy activities

If activities are to be described by their *ij* numbers, then it is essential that no two activities have the same numbers. Now this can present difficulty: consider the following extract from a project's precedence table.

Example 1

Activity	Preceding activity
B and C	A
D	B and C

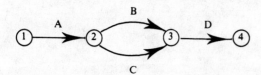

Figure 18.10

Now this network (Fig. 18.10) does show a logical sequence of activities, but we cannot accept it, as activities B and C have the same *ij* number. We overcome this problem by introducing a *dummy activity*, which is represented by a broken line (see Fig. 18.11).

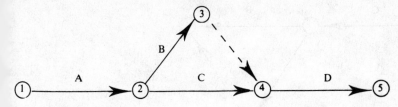

Figure 18.11

Now the dummy activity does not represent any activity as such: it is inserted to preserve the sequential numbering system of events. Such a dummy is called an *identity dummy*.

Sometimes it is necessary to insert a dummy to preserve not the sequential numbering system, but the logical sequence of events. Consider the following precedence table.

Example 2

Activity	Preceding activity
B and C	A
D	B
E	B and C
F	D and E

You should satisfy yourself that the network in Fig. 18.12 is a true representation of the logical sequence of activities, that the dummy 3–4 is essential to preserve the logical sequence.

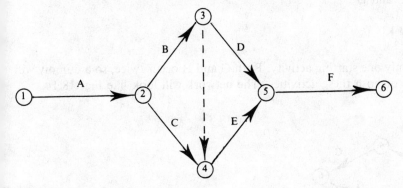

Figure 18.12

Earlier, we stated that a network has just one start event and one end event, and it is sometimes necessary to use dummy activities to ensure that this is so. The network in Fig. 18.13 illustrates this point.

It would be useful at this stage to state some rules so that we can decide when it is necessary to use dummies.

(a) If an activity occurs in the right-hand column of a precedence table but not in the left-hand column, it cannot depend upon another activity having been completed. Hence, it must be a 'starting activity'. If there is more than one start activity you *may* need dummies.

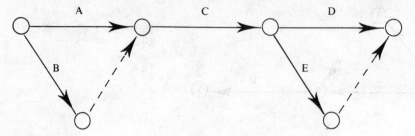

Figure 18.13

(b) If an activity occurs in the left-hand column of a precedence table but not in the right-hand column, then it must be an end activity. If there is more than one end activity you *may* need dummies.

(c) If any activity occurs more than once in the right-hand column, then you *will* need to introduce dummies. If the activity occurs n times, then $(n-1)$ dummies will have to be drawn from its end event.

The precedence table of a certain project looks like this:

Activity	Preceding activity
B, C and D	A
E and F	B
G	E
H	F
I	G and H
J	G, H, C and D
K	D
L	I, J and K

First, we note that there is only one starting activity. Both G and H occur twice, so a dummy will be needed from the end event of both these activities. The network will look like Fig. 18.14.

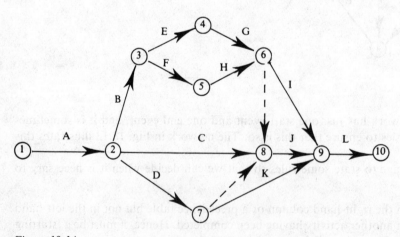

Figure 18.14

18.6 Finding the total project time

Let us suppose that the network in Fig. 18.14 describes the activities in a manufacturing process. The network clearly shows that activities B, C and D can all be started together as long as there are sufficient resources to do so. Let us assume that we can easily obtain all the resources that we need. Later, we can drop this assumption. Now let us assume that the project is such that the only resource needed is labour, and that the labour available is equally capable of performing any activity. The time taken to complete each activity is known:

Activity	A	B	C	D	E	F	G	H	I	J	K	L
Time (hrs)	3	4	5	6	2	1	7	4	3	5	6	2

How long will be the project take? To answer this, we must examine all the routes through the network. Can you see that there are seven possible routes? Let us list each and find the time taken.

Route	Time	
A B E G I L	$3+4+2+7+3+2$	= 21 hrs
A B E G Dummy J L	$3+4+2+7+0+5+2$	= 23 hrs
A B F H I L	$3+4+1+4+3+2$	= 17 hrs
A B F H Dummy J L	$3+4+1+4+0+5+2$	= 19 hrs
A C J L	$3+5+5+2$	= 15 hrs
A D Dummy I L	$3+6+0+5+2$	= 16 hrs
A D K L	$3+6+6+2$	= 17 hrs

The project cannot be completed in less than 23 hours. This is determined by the longest route through the network – called the *critical path*. Activities on this route must be completed on time otherwise the total project time will lengthen (i.e. the activities have critical times). You should realise that it would be very difficult to calculate the total time without first drawing the network. Try to find the total project time just using the precedence table and you will realise how true this is!

Most networks will be much more complicated than the one we have examined, and it will be tedious to identify all the routes. In fact, it is more than likely that some routes will be overlooked. A more efficient method is to use *the earliest event times*, i.e. the earliest time that each event can occur. We divide the circle showing the event into three parts like this:

A is the event number
B is the earliest event time

We start at 1 and arbitrarily assign to it a start time of zero. Event 2 occurs when activity A is complete, so the earliest event time is $0 + 3 =$ third hour (see Fig. 18.15).
Likewise, we can deduce that

the earliest event time for event 3 is 7 hours (activity B complete)
the earliest event time for event 4 is 9 hours (activity E complete)
the earliest event time for event 5 is 8 hours (activity F complete)
the earliest event time for event 7 is 9 hours (activity D complete)

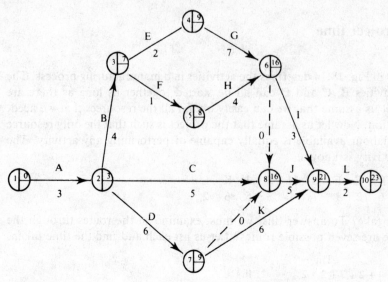

Figure 18.15

Now let us consider event 6. This cannot occur until both activity G and activity H are complete.

activity G is complete after $9 + 7 = 16$ hours at the earliest
activity H is complete after $8 + 4 = 12$ hours at the earliest

So the earliest time for event 6 is the 16th hour.

For event 8, the earliest time would be either $16 + 0 = 16$ (dummy activity 6–8 complete – remember that it is not an activity as such and so cannot occupy any time) or $3 + 5 = 8$ (activity C complete) or $9 + 0 = 9$ (dummy activity 7–8 complete). Clearly, the earliest time for event 8 is the 16th hour. Continuing in this way for all the other events, we finally reach the end event 10 which has an earliest time of 23 hours. Clearly, the earliest time for the end event must be the same as the total project time. Using this method to find the total project time is more efficient and (as we shall see later) is essential to further analysis of the network.

18.7 The critical path

Finding the earliest event times has certainly helped to determine the total project time, but it has not helped to isolate the critical path. To do this, we must use the latest event times, i.e. the latest time that each event can occur if the network is to be completed on time. We will insert the latest event times at position C in the event symbol:

Obviously, the latest time for event 10 is the 23rd hour, and the latest time for event 9 is the $23 - 2 = 21$st hour. Again, the latest time for event 8 is quite straightforward: it is $21 - 5 = 16$th hour.

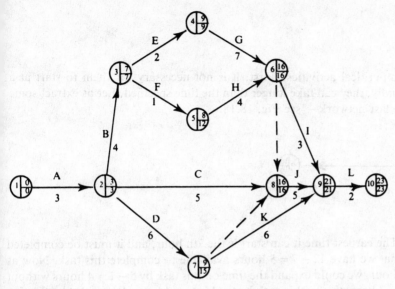

Figure 18.16

For event 6, we must consider the two following activities, 6–8 and 6–9. If activity 6–8 is to be complete on time, then event 6 has a latest time of $16 - 0 = 16$th hour. If activity 6–9 is to be complete on time, then event 6 has a latest time of $21 - 3 = 18$th hour. Can you see that if the project is to be complete on time, event 6 has the 16th hour as its latest time? Continuing in this way, we can obtain the latest times for all the events (see Fig. 18.16).

Think back to critical activities – they must be started on time, otherwise the total project time will lengthen. What does this imply? Each event on the critical path must have the same earliest and latest times. Using this fact, we can easily identify the critical path in Fig. 18.16 – but beware! The method is not infallible. Consider Fig. 18.17.

If we state that the critical path contains those events which have the same earliest and latest times, then we would isolate paths 1–2–3–5 and 1–2–5. Are they both critical paths?

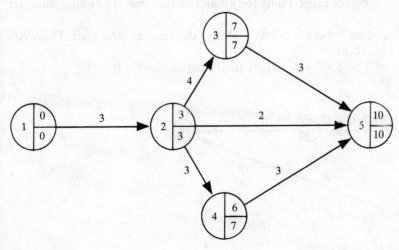

Figure 18.17

18.8 Float

What can we deduce about non-critical activities? First, it is not necessary for them to start at a particular, specified time. Secondly, they can take longer than the time specified. Let us extract some non-critical activities from the last network – see Fig. 18.18.

Figure 18.18

Consider first activity 3–5. The earliest time it can start is the 7th hour, and it must be completed by the 12th hour. It follows that we have $12 - 7 = 5$ hours available to complete this task. Now as the activity should only take 1 hour, we could expand the time on this task by $5 - 1 = 4$ hours without affecting the total project time. Alternatively, this activity could start 4 hours late without affecting the total project time. This 4 hours latitude we have on activity 3–5 is called its *total float. We can define total float as time available for an activity minus expected activity duration*, or using i as the start event and j as the end event: latest time for event j minus earliest time for event i minus activity duration.

Critical activities will have zero total float – this is the only reliable method of extracting the critical path.

The total float of activity 5–6 is $16 - 8 - 4 = 4$ hours. Now suppose we start activity 3–5 on the 11th hour (i.e. take advantage of its total float), then this task will be completed on the $11 + 1 = 12$th hour. Activity 5–6 now cannot start until the 12th hour, and its total float is $16 - 12 - 4 = 0$. Thus, the total float of 4 hours refers to the path 3–6 rather than the activities on the path.

Now suppose we take up all the float on activity 5–6, then this activity must end on the 16th hour. Now no activity starting with event 6 can begin before this time, so taking up the float on 5–6 does not affect the following activities. The float on 5–6, then, is essentially different to the float on 3–5. We say that activity 5–6 has a *free float* of 4 hours. Thus, total float affects following activities, whereas free float does not.

We can calculate free float by taking the earliest time for j minus earliest time for i minus duration of ij.

In this particular network, free float (where it exists) is always the same as total float. However, this need not be so. Consider Fig. 18.19.

Activity 1–3 has a free float of $12 - 5 - 0 = 7$, but its total float is $16 - 5 - 0 = 11$.

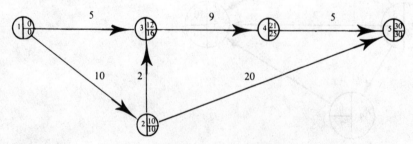

Figure 18.19

18.9 Project summary tables

Table 18.1 shows a standard format summary of our network from Fig. 18.16. The first two columns list the *ij* numbers of all the activities together with their estimated times. For any activity, we can pick up the earliest start and latest end as in Fig. 18.20.

Let us take activity F as an example (Fig. 18.21).

The earliest start is day 7, so the earliest end is day $7 + 1 = 8$. The latest end is on day 12, so the latest start is on day $12 - 1 = 11$. We can calculate total float either by subtracting earliest start from latest start, or by subtracting earliest end from latest end. Free float is calculated as in the previous section.

Table 18.1

Start event	End event	Time	Earliest		Latest		Float		Critical
			Start	End	Start	End	Total	Free	
1	2	3	0	3	0	3	0	0	*
2	3	4	3	7	3	7	0	0	*
2	7	6	3	9	9	15	6	0	
2	8	5	3	8	11	16	8	8	
3	4	2	7	9	7	9	0	0	*
3	5	1	7	8	11	12	4	0	
4	6	7	9	16	9	16	0	0	*
5	6	4	8	12	12	16	4	4	
6	8	0	16	16	16	16	0	0	*
6	9	3	16	19	18	21	2	2	
7	8	0	9	9	16	16	7	7	
7	9	6	9	15	15	21	6	6	
8	9	5	16	21	16	21	0	0	*
9	10	2	21	23	21	23	0	0	*

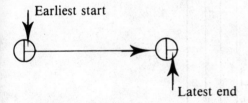

Figure 18.20

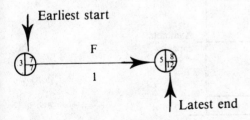

Figure 18.21

18.10 Gantt charts: 'loading' the network

So far, our analysis has ignored the supply of resources that are available to work on a project. Let us continue to analyse the same project, and assume that we require one man per activity. Allocating resources to the project is called 'loading the network', and this is usually done with a *Gantt chart*. On a Gantt chart, the activities are represented by lines having lengths proportional to the duration of each activity. The Gantt chart for our project would look like Fig. 18.22.

Each activity on the Gantt chart is identified by its *ij* number. Can you see the assumption on which the chart has been drawn? It has been assumed that each activity starts at its earliest time. The line representing activity 5–6 shows that the activity starts at the end of the 8th hour and ends at the end of the 12th hour. The dotted line 5–6 shows that activity 5–6 must be completed at the end of the 16th hour at the latest. The dotted line shows the total float of each activity. At the foot of each column, the number of men required for that hour is shown. This is obtained by counting the number

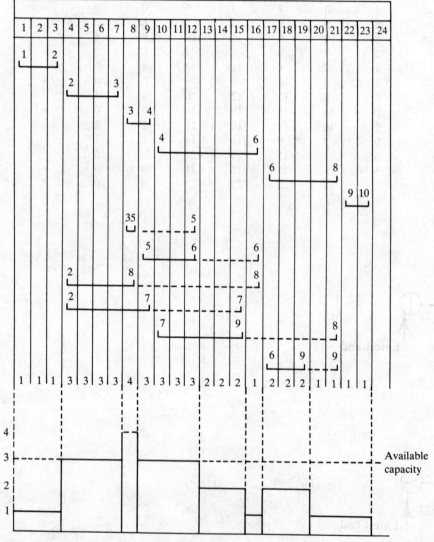

Figure 18.22

of solid horizontal lines in each column. The histogram shows the amount of labour required on an hourly basis.

The histogram shows that the labour scheduled to this project is unevenly distributed. It also shows that there are three men available for this project. For most of the time, the labour required is less than the labour available, and the project is said to be underloaded. However, in the 8th hour the project is overloaded, and this overload must be removed. It would also seem sensible to smooth the histogram as much as possible.

Obviously, we must concentrate on activities that have float, and as activities with free float do not affect other activities, we will consider them first. Activity 7–9 has a free float, and if we put this forward one hour, the gaps in the histogram in the 16th hour will be filled. This, however, will leave another gap at the 10th hour. We can fill this gap and remove the overload in the 8th hour by advancing activities 3–5 and 5–6 by two hours. This removes the overload, but does not smooth the histogram as the demand for labour is now like this:

Hour	Labour required
0–3	1 man
4–8	3 men
9–10	2 men
11–14	3 men
15–19	2 men
20–23	1 man

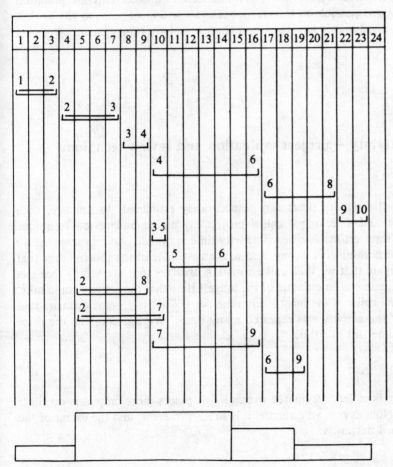

Figure 18.23

The unevenness in the 9th and 10th hours can be removed if we advance activities 2–8 and 2–7 by an hour. The Gantt chart and demand for labour now looks like Fig. 18.23.

We can now state exactly how labour will be scheduled to the project. One man can be instructed to work on the critical path. A second man will be required from the 5th to the 14th hour, working on activities 2–8, 2–5 and 5–6.

18.11   Progress charts

The Gantt chart can be used effectively to show how a project is actually progressing, and we do this by drawing a line above the activity symbol which is proportional to the amount of work completed. Suppose, for example, that the Gantt chart in Fig 18.23 represents the state of the project at the end of the 8th hour. We can conclude that activity 3–4 should be 50 per cent complete but that it has not yet been started.

Activity 2–7 should be 83.33 per cent complete – in fact, it has been completed. This interpretation begs the question as to whether it is possible to transfer labour to the critical path, as the project time is in danger of lengthening. Used in this way, the Gantt chart is an important element of project control.

18.12   Dealing with uncertainty – project evaluation and review technique (PERT)

So far, we have assumed that the activity times can be accurately estimated. In practice, such estimates will be liable to error and a technique called project evaluation and review technique (PERT) is applied. In practice, three estimates of each activity time are made. An estimate that we shall call estimate a assumes that nothing goes wrong, and so is an estimate of the shortest time that the activity could take. An estimate that we shall call estimate b assumes that everything that can go wrong does go wrong, so estimate b is an estimate of the longest time that the activity could take. Finally, an estimate that we shall call c is the most likely time for the activity – the estimate that would occur most frequently if the activity was repeated many times. We can combine the three estimates to give the expected time (te) like this:

$$te = \frac{a + b + 4c}{6}$$

An estimate made in this way is called a weighted average – it places more emphasis on the c estimate than on the other two. Now as te is an estimate, it is subject to error, and the extent of the error is measured by the standard deviation:

$$\text{standard deviation of estimate} = \frac{b - a}{6}$$

Suppose the following estimates have been made for a four-activity project:

Activity	Shortest time	Longest time	Most frequent time	Expected time	Standard deviation
A	4 days	10 days	6 days	6.33 days	1 day
B	6 days	14 days	9 days	9.33 days	1.33 days
C	5 days	8 days	6 days	6.16 days	0.5 days
D	7 days	13 days	11 days	10.67 days	1 day

If the four activities occur in sequence, then we can estimate the time for the project by adding together the individual expected times:

$$6.33 + 9.33 + 6.16 + 10.67 = 32.5 \text{ days}$$

Now we cannot add the individual standard deviations in order to obtain the standard deviation for the project, but we can add their squares:

$$\text{standard deviation for project} = \sqrt{(1)^2 + (1.33)^2 + (0.5)^2 + (1)^2} = 2.007 \text{ days}$$

If we assume that the activity times are normally distributed, then we can be more than 95 per cent sure that the project will be completed in less than:

$$\text{expected time} + 1.645 \times \text{standard deviation}$$

So we could quote a completion time for the project of $32.5 + 1.645 \times 2.007 = 35.8$ days

18.13   Decreasing the total project time

Sometimes it is possible to reduce the time spent on activities by drafting in extra resources, though of course this would increase the cost. Consider the example in Table 18.2. Here, the ij numbers are given, so drawing the network will be much easier. Normal time is the time taken if usual procedures are adopted, whereas minimum time is the time taken if extra resources are drafted in. (The cost of these extra resources is often called CRASH costs.)

Table 18.2

Task	Normal time (days)	Minimum time (days)	Cost of extra resources	Equivalent cost per day
1–2	5	4	£150	£150
2–3	9	5	£360	£90
2–4	8	6	£60	£30
3–4	0	0	0	0
3–6	4	2	£20	£10
4–5	8	7	£80	£80
5–6	0	0	0	0
5–7	4	2	£140	£70
6–7	6	3	£180	£60
7–8	2	2	0	0
			£990	

First, let us draw the network assuming normal procedures are adopted (Fig. 18.24).

Now let us assume that the extra resources are used, and insert the minimum times on the network (see Fig. 18.25).

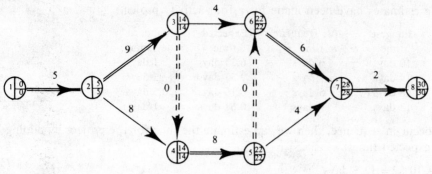

Figure 18.24

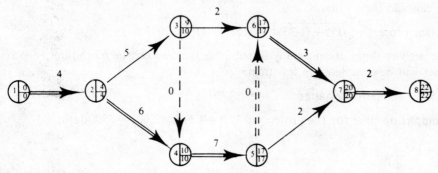

Figure 18.25

Notice that the critical path (marked by the double lines) has changed, Drafting in extra resources has reduced the project by eight days, but it has increased cost by £990. But need cost increase by this much? Surely, it is not necessary to reduce non-critical activities to a minimum! Activity 5–7 cannot start before day 17 nor finish after day 20 – it could take three days without increasing the project time. Instead, then, of decreasing activity 5–7 from four days to two days, we can decrease it from four days to three days, and assuming that the cost increase is proportional to the time reduction, we would save 140/2 = £70. We could adjust the other non-critical activities in a similar fashion. Activity 3–6 could take 17 – 9 = 8 days without affecting the total project time – it could return to its 'normal' time of four days, saving £20 on resources. Likewise, the available time for 2–3 is 10 – 4 = 6 days, and if sufficient resources are allocated to achieve this, a further £90 could be saved. A little thought, then, has reduced additional resource cost by 70 + 20 + 90 = £180, and the minimum time could be achieved for an extra outlay of 990 – 180 = £810.

The minimum time at minimum cost network shows that all activities except 3–6 are now critical (Fig. 18.26).

Sometimes it is necessary to achieve a target time, and it is desirable to achieve it at minimum cost. Suppose the target time is 25 days: to determine the best way of achieving this we will re-examine the network showing normal times, and also showing the daily cost of resources necessary to reduce the activity times (Fig. 18.27).

To achieve a project time of 25 days, we must reduce the critical path by five days. The cheapest activity to reduce is activity 6–7, which we can reduce by three days at a cost of £60 per day. However, if we were to reduce 6–7 by three days, the time for the project would not also be reduced by three days unless we also reduce activity 5–7 by one day. So to reduce the project time by three days using activity 6–7 would cost £60 per day for the first two days and £130 for the third day. For the moment we shall reduce activity 6–7, and hence the project time, by two days. The next

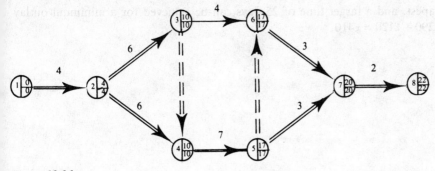

Figure 18.26

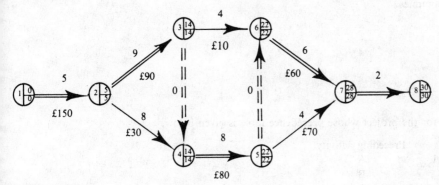

Figure 18.27

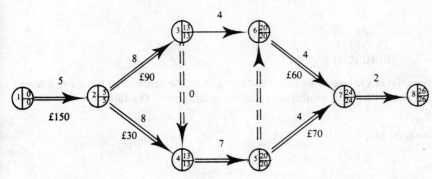

Figure 18.28

cheapest activity to reduce is activity 4–5 which can be reduced one day for an outlay of £80. We now require to reduce the critical path by a further two days. Activity 2–3 can be reduced by four days at a cost of £90 per day, but if we reduce this activity by more than one day we must also reduce activity 2–4 if the project time is also to be reduced. For the moment, then, we will reduce activity 2–3 by one day (see Fig. 18.28).

So far, we have managed to reduce the project time by four days, and this has cost us $2 \times £60 + £80 + £90 = £290$. To achieve the target time of 25 days, we must reduce the critical time by one more day. We have three ways of achieving this:

(a) reduce 1–2 at a cost of £150
(b) reduce 2–3 and 2–4 at a cost of £90 + £30 = £120
(c) reduce 5–7 and 6–7 at a cost of £70 + £60 = £130

Option (b) is the cheapest, and a target time of 25 days can be achieved for a minimum outlay on extra resources of £290 + £120 = £410.

Review questions

18.1 Draw the network of the following project for which the *ij* numbers are given for the activities. Activities 5–7 and 6–7 are dummies.

1–2	4–7
2–3	5–7
2–4	5–8
2–6	6–7
3–4	6–8
3–5	7–8

18.2 Draw the network for the project whose precedence table is given below.

Activity	Preceding activity
B, F, I, J, K, L	A
C	B
D	C
E	D
G	F
M	L
N	M
H	D, G, I, J
O	E, H, K, N

18.3 Use the networks in the solutions to questions 18.1 and 18.2 to insert the earliest and latest event times. For each project, state what is the total project time and the critical path. The times for the activities are as follows:

(a)

Activity	Duration (days)
1–2	3
2–3	4
2–4	5
2–6	6
3–4	3
3–5	2
4–7	4
5–7	0
5–8	5
6–7	0
6–8	1
7–8	3

(b)

Activity	A	B	C	D	E	F	G	H	I	J	K	L	M	N	O
Time (days)	5	10	5	2	3	8	1	12	5	5	10	2	10	6	5

Keep your answers to this question as you will need them to complete the next one.

18.4 Prepare an activity summary chart for the network you drew for question 18.3(a). Keep your answer safe as you will need it for the next question.

18.5 Prepare an activity summary chart for the network you drew for question 18.3(b).

18.6 Draw the Gantt chart for the project summary chart you derived for question 18.4, assuming that each activity starts at its earliest time, and requires one man. What is the minimum number of men necessary to complete the project in its earliest possible time?

18.7 Acme plc supplies generating equipment on rental, and is responsible for servicing the equipment. Every five years, the equipment needs a major overhaul, and engineers can predict accurately the times taken by the various tasks involved. These times can be reduced by drafting in extra resources to the tasks, and the following details are available.

Task	Normal time (days)	Normal cost	Minimum time (days)	Cost of minimum time
1–2	18	£8,000	14	£10,400
1–3	8	£6,000	6	£ 7,000
1–4	7	£7,000	5	£ 8,000
2–7	10	£6,000	6	£ 9,400
3–4	0	0	0	0
3–5	9	£8,000	6	£ 9,800
4–6	11	£9,000	7	£10,200
5–6	6	£3,000	5	£ 3,400
6–7	9	£10,000	7	£11,600

You may assume that the reduction in time and increase in cost is directly proportional.

Find the minimum time for the project and recommend a course of action to complete the project in the minimum time and at minimum cost.

Specimen examination questions

1 The table below shows, per thousand of population, the number of persons expected to survive to a given age.

Age	0	10	20	30	40	50	60	70	80	90	100
Number surviving to given age	1,000	981	966	944	912	880	748	525	261	45	0

Required:

(a) Use the table to determine the probability that
(i) a randomly chosen person born now will die before reaching the age of 60
(ii) a randomly chosen person who is aged 30 now will die before reaching the age of 60
(iii) a randomly chosen person who is aged 50 now will die before reaching the age of 60.
Comment on the order of magnitude of your three answers. (4 marks)

(b) An insurance company is planning to introduce a single life assurance policy for persons aged 50. This policy requires a single payment paid by the insured at the age of 50 so that if the insured dies within the next 10 years the dependent will receive a payment of £10,000; however, if the person survives, then the company will not make any payment. Ignoring interest on premiums and any administration costs, calculate the single premium that the company should charge to break even in the long run. (3 marks)

(c) If twelve people each take out a policy as described in (b) and the company charges each person a single premium of £2,000, find the probability that the company will make a loss. (6 marks)

(d) The above table was based on the ages of all people who died last year. Comment on the appropriateness to the company when calculating the premiums it should charge

(e)
Age	50	52	54	56	58	60
Number surviving to given age	880	866	846	822	788	748

(i) Given that a person aged 50 now dies before the age of 60, use this new information to estimate the expected age of death.
(ii) Calculate a revised value for the single premium described in part (b), taking into account the following information;
 – the expected age of death before 60 as estimated in (i)
 – a constant interest rate of 8 per cent per annum on premiums
 – an administration cost of £100 per policy
 – a cost of £200 to cover profit and commission. (5 marks)

2 The Cacus Chemical Company has recently developed a new product for which a substantial market is likely to exist in 1 year's time. Due to the highly unstable nature of the product, a new production process must be set up at a cost of £2.5 million to cope with the anticipated high

temperature reactions. This process will take 1 year to develop, but it is estimated that there is only a 0.55 probability that it will provide adequate standards of safety.

In the light of this, the company is considering the additional development of a computerised control system (CCS) which will detect and warn against dangerous reaction conditions. Research on the CCS will take one year and cost £1 million, and Cacus estimate that there is a 0.75 probability that the CCS can be developed successfully.

Development of the CCS can either begin immediately or be postponed for one year until the safety of the new process is known. If the CCS is developed immediately and the new process proves to have an adequate standard of safety, then the CCS will be unnecessary and the £1 million of expenditure will have been wasted. On the other hand, if the CCS is postponed and the new process turns out to be unsafe, a subsequently successful development of the CCS will have delayed the product by 1 year. If neither the new process nor the CCS are successful, there is no way in which the product can be safely manufactured, and the project will have to be abandoned.

If sales of the new product can commence in one year's time, it has been estimated that the discounted profit would amount to a total of £10 million before any allowance is made for depreciation on the new process or the CCS. If the launch of the new product is delayed by one year, however, the total return is expected to fall to £8.5 million due to the possibility of other manufacturers entering the market. For simplicity, you may ignore the effects of discounting on the expenditure on the CCS.

Required:

(a) Draw a decision tree to represent the various courses of action open to the company.

(7 marks)

(b) Which course of action would you recommend to the management of Cacus?

(5 marks)

(c) By how much would the probability of a successful new process development (currently estimated to be 0.55) have to change before you would alter your recommendation in (b)? Does the decision in question appear particularly sensitive to the value of this probability?

(8 marks)

3   Martin Electronic Controls Ltd is a small manufacturer of electronic components, specialising in industrial control systems. Prior to an external audit, the chief accountant of the company decided to undertake a preliminary spot check and sampled 18 of the 1,200 sales items from the previous month in order to have some general idea as to their overall value. The values in £ of these sales items are given in the following table.

82	30	98	116	80	150	200	88	70
90	160	100	86	76	90	140	76	68

(Note that using the usual terminology $\Sigma x = 1,800$ and $\Sigma x^2 = 207,200$.)

Required:

(a) On the basis of the sample values obtain a point estimate of the total value of all items. What basic assumptions are you making when finding this estimate?   (3 marks)

(b) Provide 95 per cent confidence limits for the total of all items. What further assumptions are you making when providing these confidence limits?   (6 marks)

(c) How many items should the external auditor investigate if he wishes his estimate of the total value of the 1,200 items to be within £5,000 of the true value?   (4 marks)

(d) Furthermore, the chief accountant made an inspection of 300 recent sales invoices so as to make sure that the number of minor irregularities – such as the wrong date, incorrect addition, not rubber-stamped, or not signed by the recommended person – lies within the company's error rate of 2 per cent. If nine of the 300 sample items contain minor irregularities, is there statistical evidence that the stipulated error rate has been violated? (5 marks)

(e) How many of the 300 sample items described in (d) would need to contain a minor irregularity in order for there to be statistical evidence, at the 5 per cent level of significance, that the stipulated error rate has been exceeded? (2 marks)

4 Venus Tableware is an important tableware producer in the country of Blueland. In this country, the amount of imports and exports of ceramic tableware is insignificant when compared to local production and so it can be reasonably concluded that the total Blueland production in any one year is essentially equal to the total sales. The following table shows the total Blueland production, the Venus Tableware sales, and the consequent market share that Venus Tableware commanded within Blueland for the eight-year period 1980–1987

Year	Total Blueland production (000 tonnes)	Venus Tableware sales (000 tonnes)	Venus Tableware market share (%)
1980	744	113	15.2
1981	773	108	14.0
1982	828	131	15.8
1983	900	144	16.0
1984	936	146	15.6
1985	977	157	16.1
1986	1,007	163	16.2
1987	1,066	175	16.4

Required:

(a) Plot on a scatter diagram the logarithm of total Blueland production against the year. Interpret your scatter diagram. (5 marks)

(b) Express the total Blueland production time series in the form

$$Y = ab^X$$

where Y = total Blueland production (in 000 tonnes)
X = year − 1980

Estimate (a) and (b) and interpret your value of (b). (7 marks)

(c) Use the model fitted in part (b) to predict the total Blueland production for the years 1988, 1989 and 1990. (3 marks)

(d) The chief accountant of Venus Tableware wishes to use the total Blueland production time series to forecast company sales. As Venus had increased its market share from 15.2 per cent to 16.4 per cent, he felt it might be wise to produce two sets of forecasts:

(i) a pessimistic forecast, assuming that their market share remains at 16.4 per cent
(ii) an optimistic forecast where the company's market share increases by 0.2 per cent over each of the next three years to reach 17.0 per cent by 1990.

Obtain a set of optimistic and pessimistic forecasts for Venus Tableware annual sales for the next three years. (2 marks)

(e) Outline the main shortcomings of regression-based forecasting techniques. (2 marks)

5 The Oxygon Office Supplies Company Ltd is a well established firm of paper merchants and stationers, which is open for 50 weeks of each year and specialises in retailing of general office supplies. Its many customers include financial institutions, legal establishments and insurance companies. However, steadily increasing operating costs have diminished their financial reserves which has prompted the chief accountant to recommend a reduction in overall stock levels.

Whereas in previous times it was common for the company to hold over twelve months' stock for many stock items in order to guarantee availability, pressures on liquidity seemed to demand a reduction in inventory levels.

The company's main selling item was a high quality typing paper which tended to have an erratic demand but can be assumed to be normally distributed with a mean of 800 boxes and a standard deviation of 250 boxes per week. The paper is supplied by the Tiara Paper Company at a cost of £2.50 per box. It was found that the lead time of supply of this paper recently has been very consistent at three weeks.

The annual cost of stockholding was estimated at 15 per cent of the stock item value and is based on the cost of storage and the company's cost of capital. In order to estimate the cost of a delivery of paper from Tiara the cost of making and receiving the order together with the associated accounting and stock control tasks requires a total effort of approximately twelve man hours, where the average wage rate is £160 per week for a 40 hour week.

Required:

(a) Outline the basic principles of inventory control and explain why a good inventory policy is of value to Oxygon. (6 marks)

(b) Calculate the economic ordering quantity of this stock item, together with the average length of time between replenishments. (5 marks)

(c) Determine the recommended reorder level if there is to be no more than a 1 per cent chance that a stockout will occur in any one replenishment period. (5 marks)

(d) Determine the total stockholding cost (storage plus delivery costs) per annum using the calculated values of the economic order quantity and reorder level. (4 marks)

6 The Caterpillar China Company Ltd produces a range of products, A, B, C, D and E. The following table shows the quantity of each of the required inputs necessary to produce one unit of each product, together with the weekly inputs available and the selling price per unit of each product.

Inputs		Products				Weekly Inputs
	A	B	C	D	E	
Raw materials (kg)	6.00	6.50	6.10	6.10	6.40	35,000
Forming (hours)	1.00	0.75	1.25	1.00	1.00	6,000
Firing (hours)	3.00	4.50	6.00	6.00	4.50	30,000
Packing (hours)	0.50	0.50	0.50	0.75	1.00	4,000
Selling price (£)	40	42	44	48	52	

The costs of each input are as follows:

Materials	£2.10 per kg
Forming	£3.00 per hour
Firing	£1.30 per hour
Packing	£8.00 per hour

(a) In order to maximise the weekly contribution to profit a linear programming package on a computer is to be used. Formulate this problem so that the data can be input, giving both the objective function and the constraints. State briefly any assumptions necessary for your model to be suitable. (9 marks)

(b) The output from the computer package produces the following final tableau of a Simplex solution to this problem:

A	B	C	D	E	X	S	T	U	Q
1	1.18	1.04	0.46	0	0.36	0	0	−2.29	3,357
0	−0.34	0.23	0.02	0	−0.18	1	0	0.14	321
0	1.37	2.97	2.28	0	−0.27	0	1	−2.79	9,482
0	−0.09	−0.02	0.52	1	−0.18	0	0	2.14	2,321
0	1.26	1.06	0.51	0	2.02	0	0	8.18	105,791

where A, B, C, D and E are the weekly production levels for the five products; X is the amount of raw material that falls short of the maximum available; S, T and U are the respective number of hours short of the maximum weekly input of forming time, firing time and packing time. Use this tableau to find the optimum weekly production plan for the Caterpillar China Company. Describe the implications of using this plan in terms of the unused resources and overall contribution to profit. (4 marks)

(c) In the context of this problem explain the meaning of 'the dual or shadow price of a resource'. (5 marks)

(d) There is a proposition that the company manufactures an additional product that would sell at £50 per unit. Each unit would need 6 kg of raw materials, 1 hour of forming time, 5 hours of firing time and 1 hour to pack. Is this a worthwhile proposition? (2 marks)

7 A particular project comprises 10 activities, which have the following durations and precedences.

Activity	Duration (days)	Immediately preceding activities
A	6	–
B	1	A
C	2	A
D	1	B
E	1	D
F	–	B
G	1	C
H	–	F, G
I	4	E, G
J	5	I

Activities F and H have uncertain durations which at this stage are difficult to estimate.
Required:

(a) Draw a suitable network to represent the inter-relationships between the 10 activities. (5 marks)

(b) What is the minimum time that the project could take, ignoring the effects of activities F and H? (4 marks)

(c) If the project must be completed in 19 days, what restriction does this place on the durations of activities F and H? (5 marks)

(d) After further investigation it is estimated that the expected times for activities F and H are 2 days and 1 day respectively. Furthermore, it may be assumed that the uncertainty in these two activity durations may be represented by a Poisson distribution. On the basis of this, what is the probability that the project will be completed in no more than 19 days?

8 The Crispie Cookie Company produces a range of pre-packed biscuits which are sold in 500 g packets. A biscuit is produced by baking a measured quantity of the appropriate mixture, and then a specified number of biscuits is placed into each packet. Because the baking process cannot be completely controlled, there is an unavoidable variation in the weights of the biscuits

produced. Each packet is therefore automatically weighed, and any biscuit weighing less than the required 500 g is returned for reprocessing. Whenever a packet is rejected in this way, it is estimated to cost £0.1

The company is about to start selling a new type of biscuit which has an average weight of 40 g and a standard deviation of 6 g. At full production, the weekly output of this biscuit will be 500,000 packets and the cost per packet (in £) is given by

$$0.05 + 0.01n$$

where n is the number of biscuits in each packet.

Required:

(a) If 13 biscuits are put into each packet, what will be the mean and standard deviation of the weight of a packet? (5 marks)

(b) Explain briefly why the weight of a packet of biscuits will be approximately normally distributed. What is the probability that a packet of 13 biscuits will be rejected as underweight? (5 marks)

(c) Determine the minimum cost number of biscuits per packet by calculating the average weekly production cost for packet sizes of 13, 14 and 15 biscuits. (10 marks)

9   The board of Priam Properties plc is considering the company's planned workload for the next financial year. Priam undertakes both the construction of new houses and also the repair and renovation of old buildings, and it is important for the company's long term plans to obtain a suitable mix of new houses and repair work each year. For planning purposes, the profit on new houses is taken to be 20 per cent of their total value whereas on repairs and renovations it is 25 per cent.

To avoid the problems of having to acquire large amounts of land for the building of new houses, Priam always plans for at least £4 million of repairs and renovation work each year but at the same time the company ensures that repairs and renovations do not account for more than half their total workload. Furthermore, in costing all new and repair work, 5 per cent of the total value of work is included to cover fixed overheads. It is the board's wish that the total of these overhead charges does in fact cover the company's actual fixed overheads which have been estimated at £500,000 for the coming year.

The other major consideration is the amount of skilled labour (bricklaying, plastering, plumbing, carpentry and electrical work) which is available. It has been estimated that £1 million of new house building involves 12 man years of skilled labour whereas £1 million of repair and renovation work requires 18 man years of skilled labour. The company currently employs 180 skilled workmen and it would be difficult to recruit more skilled labour in the short term.

(a) Develop a linear programming model of the planning problem facing Priam Properties. (7 marks)

(b) Using a graphical approach determine the total value of both new building work and repair and renovation work that should be undertaken next year to maximise profit. (9 marks)

(c) What increase in profit (if any) would be achieved if the company were to remove the condition of having at least £4 million of repair and renovation work each year. (4 marks)

10   A large firm of accountants and financial advisors uses a minicomputer to handle routine tasks such as calculations and data searches. Requests to perform these tasks are made on several office terminals linked to the minicomputer. The time to process any particular task is, on average, 7.5 minutes while the number of requests arriving at the computer each hour is given by the following

frequency table:

Number of tasks arriving in one hour	Frequency
0– 2	8
3– 5	34
6– 8	44
9–11	12
12–14	1
15	1

Required:

(a) Without carrying out the test, explain how you would perform a statistical test to find out whether or not tasks arrive at the minicomputer according to the Poisson distribution.
(6 marks)

(b) It is assumed that the queuing situation satisfies an M/M/1 model. Explain the conditions that are necessary for this assumption. (3 marks)

(c) Calculate the following operating characteristics of the system:
  (i)   The proportion of time the minicomputer is busy.
  (ii) The average time a task is in the system. (4 marks)

(d) Due to employee waiting time, it has been estimated that for each hour a task is in the system there is a 'task system cost' to the firm of approximately £5. An additional processor could be hired for £8,000 per year and would result in a reduction in time to process a task of, on average, 1.5 minutes. Assuming a 40-hour week and a 50 working-week year decide whether or not it is worth while for the firm to purchase this processor. For what value of the 'task system cost' would it be profitable for the company to purchase the processor? (7 marks)

11 The Dardanus Company operates four manufacturing plants at Adcaster, Belton, Comfield and Dimster. All large orders are dealt with centrally and allocated to the various plants by the planning director. Recently, three such have been received for 140,000, 80,000 and 100,000 units respectively and before he makes his allocation, the planning director obtains the following estimates of unit variable manufacturing cost from the works accountant of each plant:

Order	Adcaster	Belton	Comfield	Dimster
	Unit variable manufacturing cost (£)			
1	8	9	9	11
2	10	8	7	9
3	12	13	10	12

Any of these orders can be split and allocated to more than one plant.
Required:

(a) What is the minimum total variable manufacturing cost for these three orders?
(3 marks)

(b) If each plant is to be allocated an equal number of units, what would be the optimal allocation and the total variable manufacturing cost? (8 marks)

(c) If each plant must be allocated at least 50,000 units, determine by inspection the amount by which this would improve the total variable manufcturing cost obtained in (b). (5 marks)

(d) Explain why, in this situation, the specification of a minimum amount that must be allocated to each plant causes the optimum manufacturing cost to increase by £3 for every unit that is specified. (4 marks)

12 Romulus Products plc operates thirty-day terms for all its customers. Experience has shown that 80 per cent of all accounts are settled within one month, and 70 per cent of the remainder are settled during the second month, after the customer has been sent a standard 'overdue account' letter. Of those accounts still unpaid after 2 months, 50 per cent are settled during the third month after a 'final demand' has been sent.

Any accounts still not paid after three months are dealt with in one of two ways. If the amount owing exceeds £1,000 the company institutes legal proceedings to recover the money. Taking into account the legal costs involved, the proportion of the original sum owing which is ultimately recovered varies as follows:

Proportion recovered	Probability
0%– 40%	0.1
40%– 60%	0.3
60%– 80%	0.4
80%–100%	0.2

The process takes a further 3 months before payment is finally received.

If the amount owing is less than £1,000, the debt is sold to a debt collecting company in return for 50 per cent of the sum involved, which is obtained after a further month, i.e. at the end of month 4.

In recent months, the size of accounts issued by Romulus is shown by the following distribution:

Size of account	Probability
£0– £200	0.1
£200– £500	0.2
£500–£1,000	0.3
£1,000–£2,000	0.3
£2,000–£5,000	0.1

You may assume that there is no relationship between the size of the account, when it is settled and the proportion recovered; and that all accounts are settled on the last day of the month. The company's cost of capital is the equivalent of 1.5 per cent per month.

Required:

(a) What is the probability that, for any particular account, payment is received at the end of

    (i) the second month,
    (ii) the third month,
    (iii) the fourth month,
    (iv) the sixth month?   (6 marks)

(b) What is the present value of a new account which has £2,000 outstanding?

Monthly discount factors are:

Month	1	2	3	4	5	6
Factor	0.9852	0.9707	0.9563	0.9422	0.9283	0.9145

    (6 marks)

(c) Show how the system, as a whole, may be simulated by using the following random digits to determine the present value of two simulated accounts.

Account 1:	8	8	7	5	7
Account 2:	9	9	8	2	9

    (8 marks)

13 Flora Domestic Appliances Ltd manufactures a range of kitchen equipment which includes two models of washing machine, A and B. Anyone purchasing either model can take out an annual

maintenance contract for which the customer gets free repairs and maintenance in the event of any breakdown or faults occurring. The cost of any necessary parts is borne by the customer.

Recently, Flora has been evaluating the profitability of these maintenance agreements. To obtain a general indication of the costs involved, a sample of 12 contracts (6 for each model) was analysed, giving rise to the following information on the number of maintenance visits and the total time involved during the previous year.

Contract number	Model	Number of visits	Total time involved (h)
1	A	1	2.0
2	A	2	5.4
3	A	3	6.6
4	A	5	11.2
5	A	2	3.8
6	A	1	1.8
7	B	1	4.6
8	B	5	14.2
9	B	4	14.2
10	B	3	10.6
11	B	2	6.0
12	B	4	8.4

The company costs routine maintenance and repair work at £10 per visit plus £5 per hour for the time involved.

Required:

(a) Explain why it would be statistically invalid to use a t-test to compare the average number of visits for the two models.
(4 marks)
(b) Determine the mean and standard deviation of the total maintenance cost of each model.
(9 marks)
(c) Using an appropriate t-test, determine whether there is any significant difference between the annual maintenance costs of the two models.
(20 marks)

14 Transpennine plc operates a bus service between Manchester and Leeds, and in order to satisfy Department of Transport regulations, the buses it operates must be serviced weekly. The table below identifies the tasks that must be performed in the service, and the average time taken by each task.

Task	Time (h)
1–2	4
2–3	2
3–5	3
2–4	6
1–6	2
5–6	5
4–5	9
6–7	7
7–8	10
5–8	1

(a) Draw the network of the service project and insert the earliest and latest event times.
(8 marks)

(b) Explain what is meant by total float and free float, illustrating you answer by reference to activities 2–3 and 3–5.
(4 marks)
(c) Four of the tasks involved in the service involve a considerable degree of skill. The table

below identifies the times taken by four mechanics at each of these tasks (time in hours):

| | Mechanic | | | |
Task	A	B	C	D
3–5	1	4	3	4
4–5	2	14	7	13
6–7	3	6	6	13
7–8	5	13	12	10

Use a suitable algorithm to decide which mechanic should perform which task, and show what the effect of this allocation would be on the time taken to service a bus. (8 marks)

15   (a) Explain briefly the meaning of the term 'significantly different' in statistical hypothesis testing. (2 marks)

(b) Four production plants of Zeus Company Ltd are based at Aybridge, Beedon, Crambourne and Deepool. A random sample of employees at each of these four plants have been asked to give their views on a productivity-based wage deal that the company is proposing. The table below summarises their views:

| | Production plant | | | |
View	Aybridge	Beedon	Crambourne	Deepool
In favour	80	40	50	60
Against	35	30	40	25

Required:

(i)   A χ^2 analysis of the contingency table gave a χ^2 value of 7.34. Test the hypothesis that there is no significant difference in views between the production plants. Explain what action might have been taken to reach a clearer decision.
(4 marks)

(ii) The employees at Aybridge, Beedon, Crambourne and Deepool were also asked to indicate whether or not they were under 40 years of age. The numbers aged under 40 for the four production plants were 75, 48, 54 and 57 respectively of which 60, 28, 30 and 45 respectively were in favour of the wage proposal. Analyse the resulting 2×4 contingency table to test separately for the two age groups the hypothesis that there is no significant difference in the views between the two production plants.
(8 marks)

(iii) Estimate the overall percentage of the employees in favour of the wage deal for each group. Are these percentages significantly different? (6 marks)

Answers

Answers to review questions

Chapter 1

1.1   (a) 0.08   (b) 0.52

1.3   (a) $\frac{1}{12}$ (b) $\frac{1}{2}$ (c) $\frac{1}{4}$ (d) $\frac{1}{12}$

1.5   (b) 69.2%   (c) 50%   (d) (i) 0.128   (ii) 0.648

1.7   (a) 0.79   (b) 0.034

1.9   (a) (i) 0.276   (ii) 0.22   (b) Probabilities show drift to small car ownership   (c) 0.3852

1.11 $\frac{4}{11}$

1.13   (a) 0.4   (b) 0.48

Chapter 2

2.1   (a) 0.6899   (b) 0.9541

2.3   0.052

2.5   (a) 0.0144   (b) Poisson distribution with mean of 2.5

Chapter 3

3.1   (a) 77.45%   (b) 0.0068   (c) 0.13595   (d) 0.02275

3.3   16.1372 oz

3.5   (a) 0.02 mm   (b) £124.20   (c) £4.80

Chapter 4

4.1   (a) buy C   (b) buy B   (c) buy C

4.3   EMV (buy patent) = £110,000. Leave money invested.

4.5   EMV (information) = £143,749.6
       EMV (no information) = £129,687
       Buy information.

4.7   Buy Guided Systems Ltd, expand if early warning system replaced.
       EMV = £3.888 m

4.9   Buy market research if cost is less than £18,005.

Chapter 5

5.1   Trend = 3.758, 3.844, 3.800, 3.856, 3.942, 4.020, 4.268

5.3   Trend = 79.000, 81.125, 83.750, 87.000, 90.500, 92.750, 94.500, 96.125

Chapter 6

6.1   Regression line is $y = 4.4x + 51.33$
       Expenses when sales = 7,500 are £84,300
       Output of 7,780 will break even with price = £11

6.3   Trend line is $y = 9,560 + 2,980x$

Quarter	1	2	3	4
Prediction	4,802	6,174	10,976	5,488

6.5   Rank correlation coefficient = 0.7125

Chapter 8

8.1   (a) 25.14%   (b) 37.07%   (c) 53.28%   (d) 200 ± 5.88, 200 ± 7.7288

8.3   99 per cent confidence limits 2 ± 0.005152. Sample mean of 2.005 would be met very infrequently, so machine probably needs resetting.

8.5   95 per cent confidence limits = 0.07 ± 0.0224
       99 per cent confidence limits = 0.07 ± 0.0294

8.7   2.068%

Chapter 9

9.1   £18.50 ± 1.176

9.3   60

9.5   (a) 0.395   (b) 0.395 ± 0.0677, 0.395 ± 0.0895   (c) 2,295

9.7 $\hat{\mu} = 19.7$, $\hat{\sigma} = 0.46$

9.9   0.53 ± 0.0977

Chapter 10

10.2   (a) reject if greater than 30.
       (b) reject if less than 9 or greater than 31.

10.3   0.6879

Chapter 11

11.1 $Z = 3.54$ Significant at the 1 per cent level.

11.3 $t = 2.85$ Significant at 5 per cent but not at 1 per cent level.

11.5 $Z = 2.733$ Significant at 1 per cent level.

11.7 $Z = 3.91$ Significant at 1 per cent level.

11.9 $Z = 1.63$ Not significant.

Chapter 12

12.1 $\chi^2 = 7.884$ Not significant.

12.3 $\chi^2 = 8.369$ Significant at 5 per cent level.

12.5 $\chi^2 = 125.4$ Significant at 1 per cent level.

12.7 $\chi^2 = 21.875$ Significant at 1 per cent level.

Chapter 13

13.1   (a) $600/q + 1.5q$ (b) 20   (c) 2 weeks

13.3   Order 60 items every 11 days.

13.5 $q = 200$, cost = £1,640. $q = 500$, cost = £1,641.75. $q = 2,000$, cost = £1,717.
       Reject both discounts.

13.7   Reorder level = 49. Increase in cost = £3.40 per annum.

13.9   Production run size = 2,580 approx. Inventory cycle = 0.5 weeks.
       Cost = £581 per week.

Chapter 14

14.1   (a) 0.375   (b) 0.208 weeks   (c) 1.66

14.3   23 boxes sold. Total unsatisfied demand = 11. Closing stock at end of simulation = 2.

14.5   Total cycle length = 278 days. Holding cost = £245.60. Stockouts = 28.

14.7   Reorder level of 50, cost per day = £3.23.
       Reorder level of 100, cost per day = £3.36.

14.11 Total sales = 91. Unsatisfied demand = 6. Closing stock = 0.

Chapter 15

15.1 (a) 25
15.3 $1.33 < x < 8,$ $-1 < y < 0.66$
15.5 10 standard and 5 de luxe. Profit = £12.50.
15.7 Convert 36 trucks only. Profit = £7,200.

Chapter 16

16.1

0	0	1	-9	2	6
0	1	0	3	-1	6
1	0	0	-2	1	8
0	0	0	2	1	64

16.3

0	0	1	-1	1	7
1	0	0	$\frac{3}{7}$	$-\frac{5}{7}$	15
0	1	0	$-\frac{1}{7}$	$\frac{4}{7}$	16
0	0	0	$\frac{1}{7}$	$\frac{1}{7}$	47

16.5

$-\frac{13}{16}$	0	1	$\frac{15}{8}$	$-\frac{25}{16}$	75
$\frac{33}{24}$	1	0	$-\frac{5}{4}$	$\frac{45}{24}$	150
$\frac{3}{24}$	0	0	$\frac{5}{4}$	$\frac{15}{24}$	450

16.7

0	0	0	0	$\frac{3}{4}$	$-\frac{1}{2}$	0	$\frac{3}{4}$	1	0	1	15
0	0	0	0	$-\frac{1}{4}$	$\frac{1}{2}$	0	$\frac{7}{4}$	1	1	0	145
0	0	0	0	$-\frac{1}{2}$	0	1	$\frac{3}{2}$	1	0	0	100
1	0	0	0	0	0	0	-1	0	0	0	30
0	1	0	0	0	0	0	0	-1	0	0	40
0	0	1	0	$-\frac{1}{4}$	$\frac{1}{2}$	0	$\frac{7}{4}$	1	0	0	195
0	0	0	1	$\frac{3}{4}$	$-\frac{1}{2}$	0	$\frac{3}{4}$	1	0	0	25
0	0	0	0	$\frac{35}{4}$	$\frac{15}{2}$	0	$\frac{115}{4}$	15	0	0	12,225

16.9

0	1	0	$\frac{3}{4}$	$-\frac{3}{10}$	$-\frac{1}{5}$	$\frac{7}{20}$
1	0	0	$-\frac{1}{4}$	$\frac{1}{2}$	0	$\frac{7}{4}$
0	0	1	0	$-\frac{1}{5}$	$\frac{1}{5}$	$\frac{2}{5}$
0	0	0	$\frac{15}{2}$	67	18	583.5

Chapter 17

17.1

	2	3	5	6	7	
	43	65	31			2
	3	6	5	2	1	
			24	40	10	2
	1	7	5	3	2	
32						1
	0	1	3	0	-1	

7.3

128 77	206	224	75 38	158	128
110 9	184	190 35	116	122 51	110
90	35 26	73 27	160	97	−7
167	208 48	248	146 5	216 5	166
0	42	80	—53	12	

17.5

128 74	206	224	75 38	0 3	128
110 12	184 21	190 62	116	0	110
90	35 53	73	160	0	−39
167	208	248	146	0 48	128
0	74	80	—53	—128	

17.7

35 ⁴	ε ³	20 ²	10 ⁰	4
25 ⁵	²	³	⁴	3
10 ¹	⁵	³	⁴	1
0	−1	−2	−4	

17.9   Allocate A to iii, B to ii, C to i and D to iv

17.11   Southampton to Oxford
         Bristol to Taunton
         Birmingham to Northampton
         Nottingham to York
         Shrewsbury to Preston

17.13

Flight Commander	F
Pilot	A
Navigator	E
Engineer	C
Back-up man	D
Communications	B

Chapter 18

18.3   For question 18.1, total project time = 17 days, critical path 1, 2, 3, 4, 7, 8.
         For question 18.2, total project time = 39 days, critical path A, B, C, D, H, O.

18.5

Activity	Total float	Free float
J	12	0
F	8	0
I	12	12
K	19	0
L	11	0
E	9	9
G	8	8
M	11	0
N	11	11

18.7   Minimum time = 24 days.
         For minimum cost, 2–7 takes 10 days
                          1–4 takes 6 days

1   (a) (i)   As 748 people have survived to the age of 60,
$$P(\text{die before } 60) = 1 - 748/1{,}000 = 252/1{,}000 = 0.252$$
(ii)   Number dying between 30 and 60 = 944 − 748 = 196
$$P(\text{die before } 60 \mid \text{age 30 now}) = 196/944 = 0.2076$$
(iii)   Likewise;
$$P(\text{die before } 60 \mid \text{age 50 now}) = (880-748)/880 = 0.15$$
The conclusion is that the older the person, the greater his chances of surviving to the age of 60.

(b) As the probability that a person aged 50 dies before 60 is 0.15, expected average payout is $0.15 \times 10{,}000 = £1{,}500$. So a premium of £1,500 would break even.

(c) Twelve people pay a premium of $12 \times 2{,}000 = £24{,}000$, and as the sum assured is £10,000, the firm makes a loss if more than 3 people die, as the payout would then be £30,000. So we need to calculate the probability that three or more people die. To do this, we need to apply the binomial distribution with $p = 0.15$, $q = 0.85$ and $n = 12$:

$$P(3 \text{ or more}) = 1 - (P_0 + P_1 + P_2)$$
$$= 1 - (0.85)^{12} + {}^{12}C_1(0.85)^{11}(0.15) + {}^{12}C_2(0.85)^{10}(0.15)^2$$
$$= 1 - [0.1422 + 0.3012 + 0.2923] = 0.2643$$

(d) The table used as a basis for the calculations was based on the age of people who died in 1986. Now for people who are taking out insurance policies (in say 1987) the company will require tables of probabilities of dying before 1997. As this is extrapolating into the future, such things as trends in mortality will have to be taken into account.

(e) (i)   Expected age of death = mean age at death

Age at death	Frequency f	Mid-point x	fx
50–52	14	51	714
52–54	20	53	1,060
54–56	24	55	1,320
56–58	34	57	1,938
58–60	40	59	2,360
	132		7,392

Mean age at death = 7,392/122 = 56

(ii)   From the previous answer, we deduce that on average, payout is made after 6 years. Given an 8 per cent interest rate, we can calculate the present value of the average payout thus:

$$\text{P.V. of payout} = \frac{10{,}000}{(1.08)^6} = \frac{10{,}000}{1.5869} = £6{,}302$$

Probability of a payout is 0.15 (see part (a)), so expected discounted payout is $0.15 \times 6{,}302 = £945.3$

Adding administration, profit and commission, breakeven premium = 945.3 + 100 + 200 = £1,245.3.

2   (a)

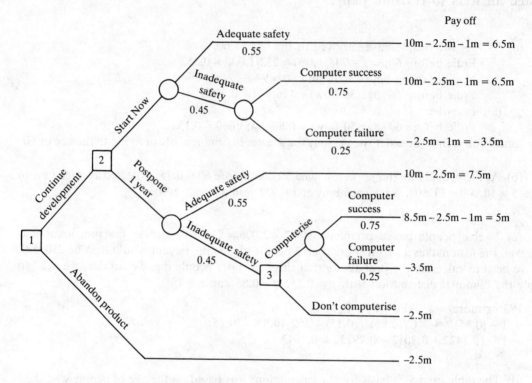

Pay off

Adequate safety
0.55
────────── 10m − 2.5m − 1m = 6.5m

Computer success
0.75
────────── 10m − 2.5m − 1m = 6.5m

Computer failure
0.25
────────── − 2.5m − 1m = −3.5m

Inadequate safety
0.45

Start Now

Postpone 1 year

Adequate safety
0.55
────────── 10m − 2.5m = 7.5m

Computer success
0.75
────────── 8.5m − 2.5m − 1m = 5m

Computer failure
0.25
────────── −3.5m

Inadequate safety
0.45

Computerise

Continue development

Abandon product

Don't computerise
────────── −2.5m

────────── −2.5m

(b) Evaluating decision point 3:
$$\text{EMV(computerise)} = 0.75 \times 5 + 0.25 \times -3.5 = £2.87 \text{ m}$$
$$\text{EMV(no computer)} = -£2.5 \text{ m}$$
So if decision point 3 reached, computerise.
Evaluating decision point 2:
$$\text{EMV(computerise now)} = 0.55 \times 6.5 + 0.45(0.75 \times 6.5 + 0.25 \times -3.5)$$
$$= £5.375 \text{ m}$$
$$\text{EMV(postpone computerisation)} = 0.55 \times 7.5 + 0.45 \times 2.875 = £5.41875 \text{ m}$$
So if decision point 2 is reached, postpone computerisation.
Evaluating decision point 1:
$$\text{EMV(develop product)} = £5.41875 \text{ m}$$
$$\text{EMV(abandon process)} = -£2.5 \text{ m}$$
So the optimal action is to continue with the process, but postpone the decision on computerisation for one year and computerise only if the process fails to meet safety standards. Note, however, that the decision to postpone the decision on computerisation is very marginal.
(c) Let p = probability that new process is successful, then
$$\text{EMV(Start now)} = 6.5p + (1 − p)(0.75 \times 6.5 + 0.25 \times -3.5)$$
$$= 6.5p + 4(1 − p)$$
$$= 2.6p + 4$$
$$\text{EMV(Postpone)} = 7.5p + 2.875(1 − p)$$
$$= 4.625p + 2.875$$
To be indifference between the two actions, EMV's must be the same, so
$$4.625p + 2.875 = 2.5p + 4$$
$$2.125p = 1.125$$
$$p = 0.53$$

If the probability that the new process is successful is less than 0.53, then we would change our optimal course of action and start the computerisation process immediately. As our estimate of success is very close to this (0.55), the decision is very sensitive to the estimate of probability.

3   (a) Point estimate of population mean = 1,800/18 = £100.
As there are 1,200 items, point estimate of total value = $1,200 \times £100 = £120,000$.
The basic assumption here is that the sample is representative of the population.

As the population is large in comparison to the sample size, we can ignore the finite population adjustment factor. Firstly, we must estimate the population variance.

$$18/17 \ [207,200/18 - 100 \times 100] = 1,600$$

Estimated population standard deviation = $\sqrt{1,600} = £40$

As n is small, we use t values with 17 degrees of freedom to set up confidence limits

$$= 100 \pm 2.11 \times 40/\sqrt{18}$$
$$= 100 \pm 19.89327$$

Estimate of total sales is $1,200[100 \pm 19.89372] = £96,128$ to $£143,872$.

The assumption here is that the population is normally distributed.

(c) We can assume that the sample will be large enough to use the normal distribution, and we require a sampling error of £5,000 at most. So using 95 per cent confidence limits

$$\frac{1.96 \times 40}{n} > \frac{5,000}{1,200}$$

$$n > \frac{1,200 \times 1.96 \times 40^2}{5,000}$$

$$n > 354$$

(d) As the distribution is obviously discrete, we must use the normal approximation to the binomial distribution. The appropriate binomial distribution is

$\pi = 0.02,$ $(1 - \pi) = 0.98,$ $n = 300$
mean $= n = 300 \times 0.02 = 6$
standard deviation $\sqrt{n\pi(1 - \pi)} = \sqrt{300 \times 0.02 \times 0.98} = 2.42487$
$H_0 = 0.02$
$H_1 \neq 0.02$

Using a normal approximation to the binomial distribution,

$$\text{Test statistic } Z = \frac{8.5 - 6}{2.42487} = 1.04$$

(remember to make the 0.5 adjustment for a discrete distribution).

For a one-sided test, the critical values of $Z = 1.645$ (at the 5 per cent level). Hence, we would accept H_0 and reject H_1. On the basis of this sample there is no evidence to suggest that the stipulated error rate has been exceeded.

(e) If the sample is significant at the 5 per cent level, then

$$\frac{(x - 0.5) - 6}{2.42487} > 1.645$$

$$x > 10.49$$

In other words, 11 irregularities would be necessary before statistical significance exists.

4   (a) First, we need to find the logarithms of total Blueland production

Total production Y	log Y
744	2.872
773	2.888
828	2.918
900	2.954
936	2.971
977	2.990
1,007	3.003
1,060	3.028

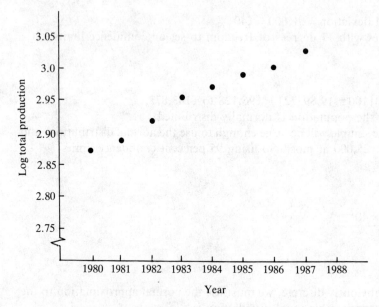

(b) If $y = ab^x$

$$\log y = \log a + x \log b$$

Hence, we can use linear regression methods to find the values of a and b in this equation. In the table below, x = year $-$ 1980 and $y = \log Y$

x	y	xy	x^2
0	2.872	0	0
1	2.888	2.888	1
2	2.918	5.836	4
3	2.954	8.862	9
4	2.971	11.884	16
5	2.990	14.950	25
6	3.003	18.018	36
7	3.028	21.196	49
28	23.624	83.634	140

$$\log b = \frac{8 \times 83.634 - 28 \times 23.624}{8 \times 140 - 28^2} = \frac{7.6}{336} = 0.0226$$

$$\log a = \frac{23.624}{8} - \frac{0.0226 \times 28}{8}$$

$$= 2.874$$

If $\log b = 0.0226$, then $b = 1.053$
If $\log a = 2.874$, then $a = 748$.

The relationship is $y = 748(1.053)^x$, showing that production is growing at 5.3 per cent per annum.

(c) 1988 = year 8, etc., so put $x = 8$, 9 and 10 in the above equation

1988	$748(1.053)^8 = 1{,}131$
1989	$748(1.053)^9 = 1{,}191$
1990	$748(1.053)^{10} = 1{,}254$

(d)

	Total production	Sales forecast Pessimistic	Optimistic
1988	1,131	$1{,}131 \times 0.164 = 185$	$1{,}131 \times 0.164 = 185$
1989	1,191	$1{,}191 \times 0.164 = 195$	$1{,}191 \times 0.166 = 200$
1990	1,254	$1{,}254 \times 0.164 = 206$	$1{,}254 \times 0.168 = 213$

The main shortcoming of regression-based forecasting techniques is the assumption that the relationships analysed are assumed to continue into the future. Generally, the longer is the period in the future forecasted, the less likely is it that the underlying assumptions will remain constant.

5   (a) The prime objective of inventory control is to minimise those costs associated with the holding of stock. Inventory costs comprise three elements. First, there is ordering cost, which is the cost associated with placing an order. Generally, the larger the batch size ordered, the smaller will be the ordering cost per time period. Second, there is holding cost which is made up of warehouse cost and the cost of capital tied up in stock. Generally, the smaller the batch size ordered the smaller will be the holding cost per time period. In its simplest form, inventory control involves balancing ordering cost per time period with holding cost per time period. Finally, there is stockout cost, which can be avoided by maintaining sufficient stock so that the probability of running out of stock is acceptably low.

It is not sensible for Oxygon to hold 12 month's supply of stock because

(i) this will tie up a considerable amount of capital which could be profitably used elsewhere;

(ii) the short lead time means that stock can be replenished quite quickly.

(b) Demand = $800 \times 50 = 40{,}000$ per annum
Cost per order = $160/40 \times 12 = £48$
Annual ordering costs = $40{,}000/q \times 48 = 1{,}920{,}000/q$
Annual holding cost = $q/2 \times 0.15 \times 2.50 = £0.1875q$

For optimum batch size
$0.1875q = 1{,}920{,}000/q$
$0.1875q^2 = 1{,}920{,}000$
$q = 3{,}200$

Number of orders = $40{,}000/3{,}200 = 12.5$
Average replenishment time = $50/12.5 = 4$ weeks
Oxygon's optimal policy is to order 3,200 boxes every 4 weeks.

(c) Lead time = 3 weeks
Average lead time demand = $3 \times 800 = 2{,}400$
Using theory of sample sums, standard error of lead time demand = $\sqrt{3 \times 250^2} = 433$.
We now must determine the lead time demand that would only be exceeded on 1 per cent of occasions.

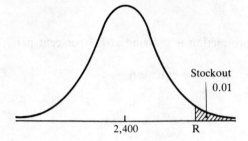

From tables, $Z = 2.33$

$$\frac{R - 2,400}{433} > 2.33$$

$$R > 2,400 + 2.33 \times 433$$
$$R = 3,409$$

Oxygon should replenish stocks when the number of boxes in stock falls to 3,409.

(d) Annual ordering costs $= 1,920,000/3,200 = £600$

Lowest stock level occurs immediately before a new batch delivered. Replenishment occurs at 3,409 units.

Lead-time demand $= 2,400$ (on average) so stock falls to $3,409 - 2,400 = 1,009$ (on average).

Highest stock level will be on delivery of new batch (of 3,200 items). So stock rises to $1,009 + 3,200 = 4,209$ (on average).

On average, stock level ranges from 1,009–4,209

Average stock held $= (1,009 + 4,209)/2 = 2,609$

Average annual holding cost $= 2,609 \times 2.5 \times 0.15 = £978.37$

Average annual cost $= 600 + 978.37 = £1,578.37$

6   (a) Firstly, calculate the revenue and cost for each product.

Product	A	B	C	D	E
Revenue	40.00	40.00	44.00	48.00	52.00
Cost	23.50	25.75	28.36	29.61	30.29
Profit	16.50	16.25	15.64	18.39	21.71

N.B. Revenue for $A = 6 \times 2.1 + 1 \times 3 + 3 \times 1.3 + 0.5 \times 8$.

Objective function:

Maximise $16.5a + 16.25b + 15.64c + 18.39d + 21.71e$

Subject to the constraints:

$$
\begin{aligned}
6a + 6.5b + 6.1c + 6.1d + 6.4e &< 35,000 \quad \text{(materials)} \\
a + 0.75b + 1.25c + d + e &< 6,000 \quad \text{(forming)} \\
3a + 4.5b + 6c + 6d + 4.5e &< 30,000 \quad \text{(firing)} \\
0.5a + 0.5b + 0.5c + 0.75d + e &< 4,000 \quad \text{(packing)}
\end{aligned}
$$

Assumptions are:

Input relationships linear
No market constraint (i.e. can sell all produced)
Constant selling price

(b) Contribution to profit will be maximised at £105,791 per week by producing 3,357 units of A and 2,321 units of E per week. Neither products B, C or D will be produced. This leaves 321 hours of forming and 9,482 units of firing surplus to requirements.

(c) The dual or shadow price of a resource is the contribution made to total profit by that resource. Thus, an extra kilo of raw material would cause profit to rise by £2.20. Likewise, an hour less of packing would cause profit to fall by £8.81. Forming and firing do not have dual values, as increases in their availability would have no affect on profit.

(d) If new product is introduced it would earn a unit profit of

$$50 - (6 \times 2.10 + 1 \times 3 + 5 \times 1.3 + 1 \times 8) = £19.90.$$

To obtain supplies of the scarce resources (raw materials and packing), less of A and E would have to be produced. This would cause a profit decline of

$$6 \times 2.02 + 1 \times 8.81 = £20.93$$

Hence, producing one unit of this new product would cause total profit to fall by £1.03, and so its production is not viable.

7   (a)

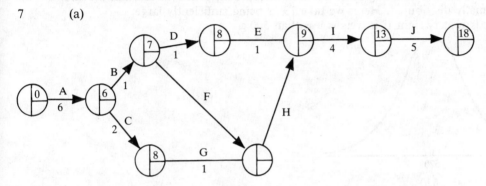

(b) 18 days

(c) Activity H occurs on the route A–C–G–H–I–J. If h is the time taken by activity H, then the time for this route is

$$6 + 2 + 1 + h + 4 + 5 = 18 + h$$

If the project must be completed in 19 days, then

$$18 + h < 19$$
$$h < 1$$

Likewise both F and H are on the route A–B–F–H–I–J.

If H takes h days, and F takes f days, then the time for this route is

$$6 + 1 + f + h + 4 + 5 = 16 + f + h$$
$$16 + f + h < 19$$
$$f + h < 3$$

So, activity H must take at most 1 day, and activities F and H together must not exceed 3 days.

(d) Given that the expected times for F and H are 2 days and 1 day respectively, the project can be completed in less than 19 days if either of the two conditions occur

H takes 0 days and F takes at most 3 days. Case 1
H takes 1 day and F takes at most 2 days. Case 2

Assuming that case 1 can exist (i.e. H can take 0 days), from tables

$$P(H = 0) = 0.368$$
$$P(F < 3) = 1 - 0.143 = 0.857$$

Using the multiplication law

$$P(H = 0 \text{ and } f < 3) = 0.368 \times 0.875 = 0.315$$

For case 2

$$P(H = 1) = 0.368$$
$$P(F < 2) = 0.857 - 0.18 = 0.677$$
$$P(H = 1 \text{ and } F < 2) = 0.368 \times 0.677 = 0.249$$

As either case 1 or case 2 would satisfy the conditions specified

$$P(\text{less than 19 days}) = 0.315 + 0.249 = 0.564.$$

8   Firstly, apply the 'sums and difference' theorem

(a) Mean weight of packet $= 13 \times 40 = 520$ g
Standard error of weights $= \sqrt{13 \times 6^2} = 21.62$ g

(b) Applying the Central Limit Theorem, if we add a large enough number of individual observations to form a sampling distribution of sample sums, then the mean of the sample sums will be normally distributed. Here, we take 13 as being sufficiently large.

A packet is underweight if it weighs less than 500 g

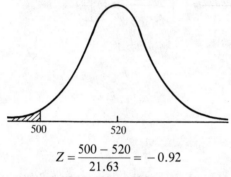

$$Z = \frac{500 - 520}{21.63} = -0.92$$

From tables, 17.9 per cent are underweight.

(c) Firstly, calculate production costs

Number of biscuits	Production cost
13	$500{,}000(0.05 + 0.01 \times 13) = $ £90,000
14	$500{,}000(0.05 + 0.01 \times 14) = $ £95,000
15	$500{,}000(0.05 + 0.01 \times 15) = $ £100,000

Now calculate the cost of rejects. For packets of 13, 17.9 per cent rejected

Reject costs $= 0.179 \times 500{,}000 \times 0.1 = $ £8,900
Total cost $\quad = $ £90,000 + £8,950 = £98,950

For packets of 14, mean $= 14 \times 40 = 560$ g,

$$\text{s.e.} = \sqrt{14 \times 6^2} = 22.45 \text{ g}$$

Percentage of rejects is given by $Z = \dfrac{500 - 560}{22.45} = 2.67$

From table, 0.4 per cent rejected

Reject costs $= 0.004 \times 500{,}000 \times 0.1 = $ £200
Total cost $\quad = $ £95,000 + £200 = 95,200

Clearly, packets of 14 are cheaper than packets of 13. As production costs alone for packets of 15 exceed total cost for packets of 14, then 14 is the optimum number per packet.

9   (a) In the usual way, we build up the model of the problem:
Let X be the value (in £m) of new houses built, and Y be the value (in £m) of renovations.

Developing the constraints in the order in which they occur in the question:
At least £4 m should be spent on renovations and repairs, so

$$Y \geqslant 4$$

Repairs and renovations should account for no more than half the total workload (which is $X + Y$) so

$$Y \leqslant 0.5(X + Y)$$
$$Y \leqslant 0.5X + 0.5Y$$
$$0.5Y \leqslant 0.5X$$
$$Y - X \leqslant 0$$

Estimated overheads = £500,000 = £0.5 m, and these should be covered by the 5 per cent of the total value of work

$$0.5(X + Y) \geqslant 0.5$$
$$X + Y \geqslant 10$$

Finally the labour used $(12X + 18Y)$ should not exceed the amount available

$$12X + 12Y \leqslant 180$$
$$2X + 3Y \leqslant 30$$

The estimated profit is $0.2X + 0.25Y$. So the problem becomes

Maximise $\pi = 0.2X + 0.25Y$
Subject to
$$Y \geqslant 4$$
$$Y - X \leqslant 0$$
$$X + Y \geqslant 10$$
$$2X + 3Y \leqslant 30$$
$$X \geqslant 0$$
$$Y \geqslant 0$$

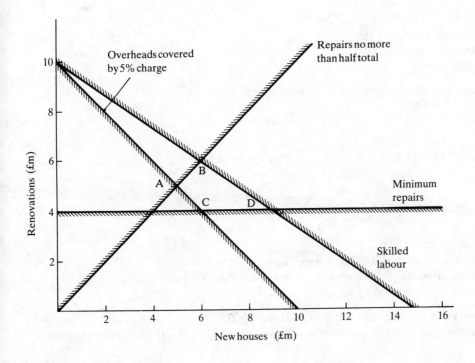

(b) Reading from the graph

Combination	X	Y	Profit £m
A	5	5	2.25
B	6	6	2.70
C	6	4	2.20
D	9	4	2.80

To maximise profit at £2.8 m, build £9 m worth of new houses and spend £4 m on renovations.

(c) Lifting the 'minimum repairs' constraint allows movement to combination E, building £15 m worth of new houses. Profit would be $15 \times 0.2 = £3$ m, an increase of £200,000.

10   (a) The appropriate test to perform here is to use χ^2. The first thing to do is to calculate the mean number of arrivals (λ) from the given frequency distribution. Next, apply the Poisson distribution

$$P(x) = \frac{e^{-1}\lambda^x}{x!}$$

where x is the number of arrivals 0, 1, 2, ..., 15 to give the probability of that number of arrivals.

Next multiply the probabilities by the total frequency, and group the frequencies as per the table given to obtain the expected (Poisson) frequencies. If any expected frequency is below 5, then combine with an adjacent class.

Now calculate the test statistic

$$\chi^2 = \Sigma \frac{(\text{observed} - \text{expected})^2}{\text{expected}}$$

and compare with the critical value of χ^2 with $n - 2$ degrees of freedom. If the calculated value of χ^2 is less than the critical value then accept the hypothesis that the tasks arrive according to a Poisson distribution.

(b) The assumptions of an M/M/1 model are:

Poisson arrivals
negative exponential service times
single queue and service point
first-come–first-served.

(c) First, calculate the mean number of arrivals per hour

Arrivals (x)	f	fx
1	8	8
4	34	136
7	44	308
10	12	120
13	1	13
15	1	15
	100	600

x is the mid-point of each class.

Mean number of arrivals $\lambda = 600/100 = 6$ per hour.

If a service takes on average 7.5 minutes, then service rate per hour $\mu = 60/7.5 = 8$.

Traffic intensity $= \lambda/\mu = 6/8 = 0.75$ hence:

(i) Proportion of time microcomputer is busy $= 0.75$

(ii) Average time spent in the system $= 1/(\mu - \lambda) = 1/(8 - 6) = 0.5$ hours $= 30$ minutes.

(d) With existing system:

Total number of arrivals $= 6 \times 40 \times 50 = 12{,}000$ per annum
Total time in system $= 12{,}000 \times 0.5 = 6{,}000$ hours
Annual task system cost $= 6{,}000 \times £5 = £30{,}000$

With second system:

Average service time is $7.5 - 1.5 = 6$ min
Service rate per hour $\mu = 60/6 = 10$ per hour
Average time spent in system $= 1/(10 - 6) = 0.25$ hour
Total time in system $= 12{,}000 \times 0.25 = 3{,}000$ hours
Annual task system cost $= 3{,}000 \times £5 + £8{,}000 = £23{,}000$

So, it is worthwhile buying a second processor as the annual saving is £7,000.
  Suppose the task system cost is £x, then for the second processor to be worthwhile,

$$3{,}000x + 8{,}000 \leqslant 6{,}000x$$
$$8{,}000 \leqslant 3{,}000x$$
$$x \geqslant £2.66$$

11   (a) Allocate order 1 to Adcaster, costing $140{,}000 \times$ £8 $= £1.12$ m
         Allocate order 2 to Comfield, costing $80{,}000 \times$ £7 $= £0.56$ m
         Allocate order 3 to Comfield, costing $100{,}000 \times £10 = £1.00$ m
         $$\underline{\text{Minimum cost} = £2.68 \text{ m}}$$

  (b) Total quantity ordered $= 320{,}000$ units so each plant will supply 80,000 units. Setting up the transportation matrix in units of 10,000 and allocating on 'least cost cell first principle' we have:

	8	9	9	11	
8	8	6	8)	8)	8
7)	10	+ 8	— 7	9	7
		ε	8	7)	
12)	12	— 13	+ 10	12	12
		2	12)	8	12
	0	1	0	0	

	8	9	9	11	
9	6	8)	10)		8
	10	8	7	9	
7)	2	6	9)		7
	12	13	10	12	
10)	11)	2	8		10
0	1	0	2		

Total cost is $8 \times £8 + 6 \times £9 + 2 \times £8 + 6 \times £7 + 2 \times £10 + 8 \times £12 = £2.92$ m.

(c) If the minimum order for each factory must be 50,000, then any particular factory may have its allocation reduced by up to 30,000 units. Take order 1. Switching 30,000 units from Belton to Adcaster would reduce costs by $30,000 \times £1 = £30,000$. Likewise, taking order 3 and switching 30,000 units from Dimster to Comfield saves $30,000 \times £2 = £60,000$. The matrix looks like this:

	8	9	9	11	
11	3	8)	10)		8
	10	8	7	9	
7)	2	6	9)		7
	12	13	10	12	
10)	10)	5	5		10
0	1	0	2		

Which is optimal, saves $£30,000 + £60,000 = £90,000$ and gives a total cost of $£2.92 - £0.09 = £2.83$ m.

(d) If no minimum quantity is specified, minimum cost = £2.68 m. If 50,000 minimum specified, minimum cost = £2.83 m.

Increase in cost due to minimum specification = £150,000, which gives a unit cost increase of $150,000/50,000 = £3$.

12 A tree diagram will help enormously here.

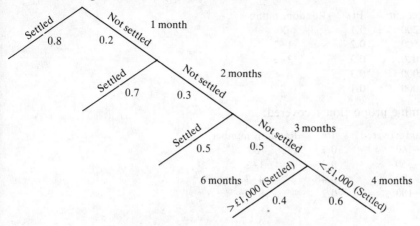

(a) P(settled after 2 months) $= 0.2 \times 0.7 = 0.14$
P(settled after 3 months) $= 0.2 \times 0.3 \times 0.5 = 0.03$
P(settled after 4 months) $= 0.2 \times 0.3 \times 0.5 \times 0.6 = 0.018$
P(settled after 6 months) $= 0.2 \times 0.3 \times 0.5 \times 0.4 = 0.012$

(b) The sum outstanding is £2,000, so payment will be received after 1, 2, 3 or 6 months. We now need to calculate the expected proportion recovered.

Proportion recovered	Pr	
x	f	fx
20	0.1	2
50	0.3	15
70	0.4	28
90	0.2	18
	1.0	63

So we expect 63 per cent to be recovered, so the amount recovered is $2,000 \times 0.63 = £1,200$. Using the discount table provided, we can now calculate the present value of the sum received.

Account settled after month	Pr	Discount factor	Amount	Present value
1	0.80	0.9852	2,000	1,576.32
2	0.14	0.9797	2,000	271.80
3	0.03	0.9563	2,000	57.38
4	0.03	0.9145	1,260	34.59
				1,940.09

N.B. If account not settled in the first three months, it *must* be settled in the 6th month. Probability of this =

$1 - (0.8 + 0.14 + 0.03) = 0.03$.

(c) Firstly, allocate random numbers to the distributions:
(i) To determine when account settled:

Month	Pr	Random number
1	0.08	00–79
2	0.14	80–93
3	0.03	94–96
>3	0.03	97–99

(ii) To determine size of account:

Size of account	Pr	Random number
0– 200	0.1	00
200– 500	0.2	1–2
500–1000	0.3	3–5
1000–2000	0.3	6–8
2000–5000	0.1	9

(iii) To determine proportion recovered:

Percentage recovered	Pr	Random number
0– 40	0.1	00
40– 60	0.3	1–3
60– 80	0.4	4–7
80–100	0.2	8–9

For account 1:

Determine month settled: Random number = 88, so settled after 2 months
Determine size of account: Random number = 7, size = £1,500 (average)
Present value = $1,500 \times 0.9707 = £1,456.05$.

For account 2:

Random number = 99, settled after 3 months
Random number = 8, size of account = £1,500 (average)

Account not sold to collection agency, so determine proportion recovered: Random number = 2, 50 per cent recovered

Present value = $1,500 \times 0.5 \times 0.9154 = £685.88$.

13   (a) The assumption behind a t-test is that the samples have been drawn from a normal population. However, the number of visits form a discrete distribution, and by examination the distributions are skewed.

(b) Beware! Question asks for estimate of population mean and variance.
Taking model A:

Visits	Time	Cost = $10v + 5t$	$(Cost)^2$
1	2.0	20	400
2	5.4	47	2,209
3	6.6	63	3,969
5	11.2	106	11,236
2	3.8	39	1,521
1	1.8	19	361
		294	19,696

Mean = 294/6 = £49
$$S^2 = 19,696/6 - 49^2 = £881.66$$

$$S^2 = \sqrt{\frac{881.66 \times 6}{5}} = £32.53$$

Taking model B:

Visits	Time	Cost = $10v + 5t$	(Cost)^2
1	4.6	33	1,089
5	14.2	121	14,641
4	14.2	111	12,321
3	10.6	83	6,889
2	6.0	50	2,500
4	8.4	82	6,724
		480	44,164

Mean = $480/6$ = £80

$S^2 = 44,164/6 - 80^2 = £960.66$

$$\hat{S}^2 = \sqrt{\frac{960.66 \times 6}{5}} = £33.95$$

(c) The question asks for a two-sided test

H_0: $\mu_1 = \mu_2$
H_1: $\mu_1 < > \mu_2$

Remember that $n/(n-1)$ correction already made!!

Standard error of mean for sample A = $32.53/\sqrt{6}$ = 13.28
Standard error of mean for sample B = $33.95/\sqrt{6}$ = 13.86

Standard error of difference = $\sqrt{13.28^2 + 13.28^2}$ = 19.2

Test statistic $t = \dfrac{|80 - 49| - 0}{19.2} = 1.61$

For a two-sided test with $6 + 6 - 2 = 10$ degrees of freedom, critical value of $t = 2.228$ (5 per cent level).
Accept H_0 and reject H_1.
No evidence to suggest any difference between average maintenance costs.

14   (a)

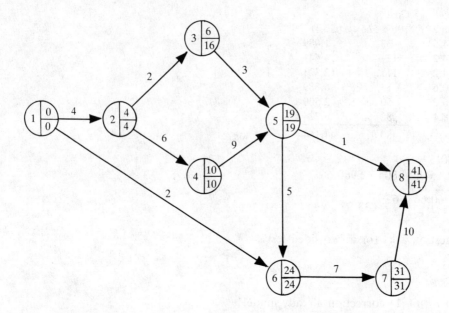

(b) Total float is the slack in time which can be used to lengthen the activity time or start the activity late without affecting the total project time. However, using total float may affect following activities. Critical activities have a zero total float. Free float is similar but does not affect any following activities.

Total float (i, j) = latest time for j minus earliest time for i minus duration i, j.

Free float (i, j) = earliest time for j minus earliest time for i minus duration i, j.

Activity	Total float	Free float
2–3	10	0
3–5	10	10

(c)

1	4	3	4	1
2	14	7	13	2
3	6	6	13	3
5	13	12	10	5

Subtract row minima:

0	3	2	3
0	12	5	11
0	3	3	10
0	8	7	5
0	3	2	3

Subtract column minima:

0	0	0	0
0	9	3	8
0	0	1	7
0	5	5	2

Smallest uncovered element = 1:

1	1	0	0
0	9	2	7
0	0	0	6
0	5	4	1

Smallest uncovered element = 1:

2	1	0	0
0	8	1	6
1	0	0	6
0	4	3	2

Task	Mechanic	Time
3–5	C	3
4–5	A	2
6–7	B	6
7–8	D	10

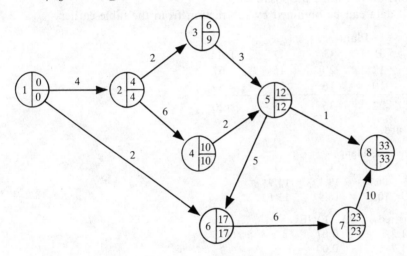

15   (a) If the difference between the observed sample results and expected sample results are unlikely to have occurred by chance, then the sample results are statistically significant.

(b) (i) Let us adopt as the null hypothesis that there is no significant difference in opinion in the three plants. Test statistic $\chi^2 = 7.24$, and with 3 degrees of freedom the critical value is 7.8 (at the 5 per cent level). Here, we have a borderline case and a larger sample should be drawn.

(ii)

	Plant				
	A	B	C	D	Total
In favour	60	28	30	45	163
Against	15	20	24	12	71
Total	75	48	54	57	234

Expected number in favour at A = $163 \times 75/234 = 52.2$
Expected number in favour at B = $163 \times 48/234 = 33.4$
Expected number in favour at C = $163 \times 54/234 = 37.6$
Remaining values can be obtained by subtraction.

	Plant			
	A	B	C	D
In favour	55.2	33.4	37.6	39.8
Against	22.8	14.6	16.4	17.2

Observed	Expected	$(O - E)^2/E$
60	52.2	1.17
28	33.4	0.87
30	37.6	1.54
45	39.8	0.68
15	22.8	2.67
20	14.6	2.00
24	16.4	3.52
12	17.2	1.57

Test statistic $\chi^2 = \quad\quad 14.02$

Critical value at the 1 per cent level is 11.3, so there is strong evidence to suggest the plants have different opinions on the wage proposal.

The 'not under 40' data can be obtained by subtraction from the table earlier

| | Plant | | | | |
	A	B	C	D	Total
In favour	20	12	20	15	67
Against	20	10	16	13	59
Total	40	22	36	28	126

The expected values are:

| | Plant | | | |
	A	B	C	D
In favour	21.2	11.7	19.1	14.9
Against	18.8	10.3	16.9	13.1

Observed	Expected	$(O - E)^2/E$
20	21.2	0.07
12	11.7	0.00
20	19.1	0.04
15	14.9	0.00
20	18.8	0.08
10	10.3	0.00
16	16.9	0.05
13	13.1	0.00
		0.24

Here, we would conclude that the proportion under 40 years of age in each plant in favour of the proposal could well be the same in the four plants.

(iii) Proportion in favour under 40 = $163/234 = 0.697$
Proportion in favour over 40 = $67/126 = 0.532$

$H_0: \pi_1 = \pi_2$
$H_0: \pi_1 <> \pi_2$

If H_0 is true, estimated proportion in combined population =

$$\frac{163 + 67}{234 + 126} = \frac{230}{360}$$

Estimated standard error of difference between proportions

$$\sqrt{230/360 \times 130/360 \times (1/234 + 1/126)} = 0.053$$

Test statistic $Z = \dfrac{0.697 - 0.532}{0.053} = 3.1$

For a two-sided test, critical value of $Z = 2.576$ (1 per cent level). Reject H_0 and accept H_1. Significant difference between proportions established.

Tables

Values of e^{-x} *(for use with Poisson Distribution)*

x	.00	.01	.02	.03	.04	.05	.06	.07	.08	.09
0.0	1.0000	.9900	.9802	.9704	.9608	.9512	.9418	.9324	.9231	.9139
0.1	.9048	.8958	.8869	.8781	.8694	.8607	.8521	.8437	.8353	.8270
.2	.8187	.8106	.8025	.7945	.7866	.7788	.7711	.7634	.7558	.7483
.3	.7408	.7334	.7261	.7189	.7118	.7047	.6977	.6907	.6839	.6771
.4	.6703	.6637	.6570	.6505	.6440	.6376	.6313	.6250	.6188	.6126
.5	.6065	.6005	.5945	.5886	.5827	.5769	.5712	.5655	.5599	.5543
.6	.5488	.5434	.5379	.5326	.5273	.5220	.5169	.5117	.5066	.5016
.7	.4966	.4916	.4868	.4819	.4771	.4724	.4677	.4630	.4584	.4538
.8	.4493	.4449	.4404	.4360	.4317	.4274	.4232	.4190	.4148	.4107
.9	.4066	.4025	.3985	.3946	.3906	.3867	.3829	.3791	.3753	.3716
1.0	0.3679	.3642	.3606	.3570	.3535	.3499	.3465	.3430	.3396	.3362
1.1	.3329	.3296	.3263	.3230	.3198	.3166	.3135	.3104	.3073	.3042
.2	.3012	.2982	.2952	.2923	.2894	.2865	.2837	.2808	.2780	.2753
.3	.2725	.2698	.2671	.2645	.2618	.2592	.2567	.2541	.2516	.2491
.4	.2466	.2441	.2417	.2393	.2369	.2346	.2322	.2299	.2276	.2254
.5	.2231	.2209	.2187	.2165	.2144	.2122	.2101	.2080	.2060	.2039
.6	.2019	.1999	.1979	.1959	.1940	.1920	.1901	.1882	.1864	.1845
.7	.1827	.1809	.1791	.1773	.1755	.1738	.1720	.1703	.1686	.1670
.8	.1653	.1637	.1620	.1604	.1588	.1572	.1557	.1541	.1526	.1511
.9	.1496	.1481	.1466	.1451	.1437	.1423	.1409	.1395	.1381	.1367
2.0	0.1353	.1340	.1327	.1313	.1300	.1287	.1275	.1262	.1249	.1237
2.1	.1225	.1212	.1200	.1188	.1177	.1165	.1153	.1142	.1130	.1119
.2	.1108	.1097	.1086	.1075	.1065	.1054	.1044	.1033	.1023	.1013
.3	.1003	.0993	.0983	.0973	.0963	.0954	.0944	.0935	.0925	.0916
.4	.0907	.0898	.0889	.0880	.0872	.0863	.0854	.0846	.0837	.0829
.5	.0821	.0813	.0805	.0797	.0789	.0781	.0773	.0765	.0758	.0750
.6	.0743	.0735	.0728	.0721	.0714	.0707	.0699	.0693	.0686	.0679
.7	.0672	.0665	.0659	.0652	.0646	.0639	.0633	.0627	.0620	.0614
.8	.0608	.0602	.0596	.0590	.0584	.0578	.0573	.0567	.0561	.0556
.9	.0550	.0545	.0539	.0534	.0529	.0523	.0518	.0513	.0508	.0503
3.0	0.0498	.0493	.0488	.0483	.0478	.0474	.0469	.0464	.0460	.0455
3.1	.0450	.0446	.0442	.0437	.0433	.0429	.0424	.0420	.0416	.0412
.2	.0408	.0404	.0400	.0396	.0392	.0388	.0384	.0380	.0376	.0373
.3	.0369	.0365	.0362	.0358	.0354	.0351	.0347	.0344	.0340	.0337
.4	.0334	.0330	.0327	.0324	.0321	.0317	.0314	.0311	.0308	.0305
.5	.0302	.0299	.0296	.0293	.0290	.0287	.0284	.0282	.0279	.0276
.6	.0273	.0271	.0268	.0265	.0260	.0257	.0257	.0255	.0252	.0250
.7	.0247	.0245	.0242	.0240	.0238	.0235	.0233	.0231	.0228	.0226
.8	.0224	.0221	.0219	.0217	.0215	.0213	.0211	.0209	.0207	.0204
.9	.0202	.0200	.0198	.0196	.0194	.0193	.0191	.0189	.0187	.0185
4.0	0.0183									
x	.00	.01	.02	.03	.04	.05	.06	.07	.08	.09

Areas in the right-hand tail of the normal distribution

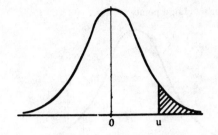

Z	0.00	0.01	0.02	0.03	0.04	0.05	0.06	0.07	0.08	0.09
0.0	.5000	.4960	.4920	.4880	.4840	.4801	.4761	.4721	.4681	.4641
0.1	.4602	.4562	.4522	.4483	.4443	.0404	.4364	.4325	.4286	.4247
0.2	.4207	.4168	.4129	.4090	.4052	.4013	.3974	.3936	.3897	.3859
0.3	.3821	.3783	.3745	.3707	.3669	.3632	.3594	.3557	.3520	.3483
0.4	.3446	.3409	.3372	.3336	.3300	.3264	.3228	.3192	.3156	.3121
.5	.3085	.3050	.3015	.2981	.2946	.2912	.2877	.2843	.2810	.2776
0.6	.2743	.2709	.2676	.2643	.2611	.2578	.2546	.2514	.2483	.2451
0.7	.2420	.2389	.2358	.2327	.2296	.2266	.2236	.2206	.2177	.2148
0.8	.2119	.2090	.2061	.2033	.2005	.1977	.1949	.1922	.1894	.1867
0.9	.1841	.1814	.1788	.1762	.1736	.1711	.1685	.1660	.1635	.1611
1.0	.1587	.1562	.1539	.1515	.1492	.1469	.1446	.1423	.1401	.1379
1.1	.1357	.1335	.1314	.1292	.1271	.1251	.1230	.1210	.1190	.1170
1.2	.1151	.1131	.1112	.1093	.1075	.1056	.1038	.1020	.1003	.0985
1.3	.0968	.0951	.0934	.0918	.0901	.0885	.0869	.0853	.0838	.0823
1.4	.0808	.0793	.0778	.0764	.0749	.0735	.0721	.0708	.0694	.0681
1.5	.0668	.0655	.0643	.0630	.0618	.0606	.0594	.0582	.0571	.0559
1.6	.0548	.0537	.0526	.0516	.0505	.0495	.0485	.0475	.0465	.0455
1.7	.0446	.0436	.0427	.0418	.0409	.0401	.0392	.0384	.0375	.0367
1.8	.0359	.0351	.0344	.0336	.0329	.0322	.0314	.0307	.0301	.0294
1.9	.0287	.0281	.0274	.0268	.0262	.0256	.0250	.0244	.0239	.0233
2.0	.02275	.02222	.02169	.02118	.02068	.02018	.01970	.01923	.01876	.01831
2.1	.01786	.01743	.01700	.01659	.01618	.01578	.01539	.01500	.01463	.01426
2.2	.01390	.01355	.01321	.01287	.01255	.01222	.01191	.01160	.01130	.01101
2.3	.01072	.01044	.01017	.00990	.00964	.00939	.00914	.00889	.00866	.00842
2.4	.00820	.00798	.00776	.00755	.00734	.00714	.00695	.00676	.00657	.00639
2.5	.00621	.00604	.00587	.00570	.00554	.00539	.00523	.00508	.00494	.00480
2.6	.00466	.00453	.00440	.00427	.00415	.00402	.00391	.00379	.00368	.00357
2.7	.00347	.00336	.00326	.00317	.00307	.00298	.00289	.00280	.00272	.00264
2.8	.00256	.00248	.00240	.00233	.00226	.00219	.00212	.00205	.00199	.00193
2.9	.00187	.00181	.00175	.00169	.00164	.00159	.00154	.00149	.00144	.00139
3.0	.00135									
3.1	.00097									
3.2	.00069									
3.3	.00048									
3.4	.00034									
3.5	.00023									
3.6	.00016									
3.7	.00011									
3.8	.00007									
3.9	.00005									
4.0	.00003									

Student's t critical points

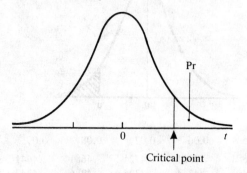

Pr

0

Critical point

t

Pr	0.25	0.10	0.05	0.025	0.010	0.005	0.001
d.f.							
1	1.000	3.078	6.314	12.706	31.821	63.657	318.31
2	0.816	1.886	2.920	4.303	6.965	9.925	22.326
3	0.765	1.638	2.353	3.182	4.541	5.841	10.213
4	0.741	1.533	2.132	2.776	3.747	4.604	7.173
5	0.727	1.476	2.015	2.571	3.365	4.032	5.893
6	0.718	1.440	1.943	2.447	3.143	3.707	5.208
7	0.711	1.415	1.895	2.365	2.998	3.499	4.785
8	0.706	1.397	1.860	2.306	2.896	3.355	4.501
9	0.703	1.383	1.833	2.262	2.821	3.250	4.297
10	0.700	1.372	1.812	2.228	2.764	3.169	4.144
11	0.697	1.363	1.796	2.201	2.718	3.106	4.025
12	0.695	1.356	1.782	2.179	2.681	3.055	3.930
13	0.694	1.350	1.771	2.160	2.650	3.012	3.852
14	0.692	1.345	1.761	2.145	2.624	2.977	3.787
15	0.691	1.341	1.753	2.131	2.602	2.947	3.733
16	0.690	1.337	1.746	2.120	2.583	2.921	3.686
17	0.689	1.333	1.740	2.110	2.567	2.898	3.646
18	0.688	1.330	1.734	2.101	2.552	2.878	3.610
19	0.688	1.328	1.729	2.093	2.539	2.861	3.579
20	0.687	1.325	1.725	2.086	2.528	2.845	3.552
21	0.686	1.323	1.721	2.080	2.518	2.831	3.527
22	0.686	1.321	1.717	2.074	2.508	2.819	3.505
23	0.685	1.319	1.714	2.069	2.500	2.807	3.485
24	0.685	1.318	1.711	2.064	2.492	2.797	3.467
25	0.684	1.316	1.708	2.060	2.485	2.787	3.450
26	0.684	1.315	1.706	2.056	2.479	2.779	3.435
27	0.684	1.314	1.703	2.052	2.473	2.771	3.421
28	0.683	1.313	1.701	2.048	2.467	2.763	3.408
29	0.683	1.311	1.699	2.045	2.462	2.756	3.396
30	0.683	1.310	1.697	2.042	2.457	2.750	3.385
40	0.681	1.303	1.684	2.021	2.423	2.704	3.307
60	0.679	1.296	1.671	2.000	2.390	2.660	3.232
120	0.677	1.289	1.658	1.980	2.358	2.617	3.160
∞	0.674	1.282	1.645	1.960	2.326	2.576	3.090

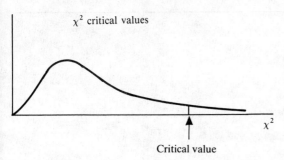

χ² critical values

Critical value

Pr d.f.	0.250	0.100	0.050	0.025	0.010	0.005	0.001
1	1.32	2.71	3.84	5.02	6.63	7.88	10.8
2	2.77	4.61	5.99	7.38	9.21	10.6	13.8
3	4.11	6.25	7.81	9.35	11.3	12.8	16.3
4	5.39	7.78	9.49	11.1	13.3	14.9	18.5
5	6.63	9.24	11.1	12.8	15.1	16.7	20.5
6	7.84	10.6	12.6	14.4	16.8	18.5	22.5
7	9.04	12.0	14.1	16.0	18.5	20.3	24.3
8	10.2	13.4	15.5	17.5	20.1	22.0	26.1
9	11.4	14.7	16.9	19.0	21.7	23.6	27.9
10	12.5	16.0	18.3	20.5	23.2	25.2	29.6
11	13.7	17.3	19.7	21.9	24.7	26.8	31.3
12	14.8	18.5	21.0	23.3	26.2	28.3	32.9
13	16.0	19.8	22.4	24.7	27.7	29.8	34.5
14	17.1	21.1	23.7	26.1	29.1	31.3	36.1
15	18.2	22.3	25.0	27.5	30.6	32.8	37.7
16	19.4	23.5	26.3	28.8	32.0	34.3	39.3
17	20.5	24.8	27.6	30.2	33.4	35.7	40.8
18	21.6	26.0	28.9	31.5	34.8	37.2	42.3
19	22.7	27.2	30.1	32.9	36.2	38.6	43.8
20	23.8	28.4	31.4	34.2	37.6	40.0	45.3
21	24.9	29.6	32.7	35.5	38.9	41.4	46.8
22	26.0	30.8	33.9	36.8	40.3	42.8	48.3
23	27.1	32.0	35.2	38.1	41.6	44.2	49.7
24	28.2	33.2	36.4	39.4	43.0	45.6	51.2
25	29.3	34.4	37.7	40.6	44.3	46.9	52.6
26	30.4	35.6	38.9	41.9	45.6	48.3	54.1
27	31.5	36.7	40.1	43.2	47.0	49.6	55.5
28	32.6	37.9	41.3	44.5	48.3	51.0	56.9
29	33.7	39.1	42.6	45.7	49.6	52.3	58.3
30	34.8	40.3	43.8	47.0	50.9	53.7	59.7
40	45.6	51.8	55.8	59.3	63.7	66.8	73.4
50	56.3	63.2	67.5	71.4	76.2	79.5	86.7
60	67.0	74.4	79.1	83.3	88.4	92.0	99.6
70	77.6	85.5	90.5	95.0	100	104	112
80	88.1	96.6	102	107	112	116	125
90	98.6	108	113	118	124	128	137
100	109	118	124	130	136	140	149

Index

additive model, 72–73
alternative hypothesis, 130
artificial variables, 213–4
assignment, 242–246

Bayes' theorem, 17
Bessel's correction, 117
binomial distribution, 25–9
    mean and variance, 28–9
    normal approximation, 40–1
bivariate distributions, 75–6
buffer stocks, 169

central limit theorem, 104
chi-square test, 148–58
cluster sampling, 101
coefficient of determination, 88
collectively exhaustive, 4
combinations, 27–8
conditional probability, 14–7
confidence limits, 109–110
contingency tables, 153–6
correlation, 85–92
correlation coefficient, 88
cost minimisation, 215–9
crash costs, 265
critical path, 258–9
curve fitting, 83–85

degeneracy, 233–7
degrees of freedom, 120–1
deseasonalisation, 67–9
detached coefficient, 204
dispatch costs, 233–7
dual solution, 216–8
dummy activities, 254–6

economic ordering quantity, 162
estimation, 113–23
    of a population mean, 113–5
    of a population proportion, 115–6
    of a population variance, 117–20
event, 3
expected frequency, 149–51
expected monetary value, 47–8

finite population adjustment, 121–2
forecasting, 70–1
free float, 260

Gantt chart, 262–4
Gosset, W.S., 118

Hungarian method, 243–6
hypothesis testing, 124–32

identity dummy, 255
inequalities, 189–93
interval estimate, 113
inventories   171
inventory costs, 161–2

lead time, 168
least squares line, 79–81
level of significance, 130
linear programming, 193–201
logical dummy, 255

M/M/1 model, 173–5
maximax criterion, 44
maximin criterion, 45
minimax criterion, 45
Monte Carlo methods, 177
moving average, 63–9
multiplicative model, 72–3